Control Engineering

Series Editor

William S. Levine
Department of Electrical and Computer Engineering
University of Maryland
College Park, MD 20742-3285
USA

Editorial Advisory Board

Panagiotis D. Christofides,
Antonios Armaou
Yiming Lou
Amit Varshney

Control and Optimization of Multiscale Process Systems

Birkhäuser
Boston • Basel • Berlin

Panagiotis D. Christofides
University of California, Los Angeles
Department of Chemical & Biomolecular
    Engineering
Los Angeles
USA
pdc@seas.ucla.edu

Antonios Armaou
Pennsylvania State University
Department of Chemical Engineering
University Park
USA
armaou@psu.edu

Yiming Lou
United Technologies Corporation
Pomona, CA
USA
ylou@ieee.org

Amit Varshney
Westminster, CO
USA
amitvarshney00@yahoo.com

ISBN: 978-0-8176-4792-6   e-ISBN: 978-0-8176-4793-3
DOI 10.1007/978-0-8176-4793-3

Library of Congress Control Number: 2008939376

Printed on acid-free paper

birkhauser.com

Contents

List of Figures

List of Tables

Preface

Multiscale process systems are characterized by highly coupled phenomena that occur in disparate spatial and temporal scales. Examples include the chemical vapor deposition of thin films, as well as ion-sputtering and catalytic processes where gas-phase and surface processes strongly interact. Detailed modeling of multiscale process systems naturally leads to continuum laws for the macroscopic (gas-phase) phenomena coupled with stochastic simulations for the microscopic (surface) phenomena. Control and optimization of multiscale process systems, targeting regulation of microscopic properties like thin-film surface roughness, cannot be addressed using existing methods that rely on continuum process models in the form of linear/nonlinear differential equations.

This book—the first of its kind—presents general, yet practical, methods for model-based feedback control and optimization of multiscale process systems. Beginning with an introduction to general issues on control and optimization of multiscale processes and a review of previous work in this area, the book discusses detailed modeling approaches for multiscale processes with emphasis on the theory and implementation of kinetic Monte Carlo simulation, methods for feedback control using kinetic Monte Carlo models, stochastic model construction and parameter estimation, predictive and covariance control using stochastic partial differential equation models, and both steady-state and dynamic optimization algorithms that efficiently address coupled macroscopic and microscopic objectives. The methods are applied to various multiscale/microscopic processes—including thin-film deposition processes, an ion-sputtering process, and a catalytic CO oxidation process—and their effectiveness and performance are evaluated through detailed computer simulations. The book also includes discussions of practical implementation issues that can help researchers and engineers understand the development and application of the methods in greater depth.

The book assumes a basic knowledge about differential equations, probability theory, and control theory and is intended for researchers, graduate students, and process control engineers.

In addition to our work, Dr. Dong Ni and doctoral candidate Gangshi Hu at UCLA contributed greatly to the research results included in the book and in the preparation of the final manuscript. We would like to thank them for their hard work and contributions. We would also like to thank all the other people who contributed in some way to this project. In particular, we would like to thank our colleagues at UCLA and Penn State for creating a pleasant working environment, the staff of Birkhäuser for excellent cooperation, and the United States National Science Foundation for financial support. Last, but not least, we would like to express our deepest gratitude to our families for their dedication, encouragement, and support over the course of this project. We dedicate this book to them.

Panagiotis D. Christofides, Antonios Armaou, Yiming Lou, and Amit Varshney

September 2008

1

Introduction

1.1 Motivation

Over the last 10 years, increasingly tight product quality specifications have motivated extensive research on the development of control and optimization methods for distributed and multiscale process systems using increasingly detailed process descriptions. On one hand, for distributed process systems for which continuum laws are applicable, nonlinear distributed parameter systems, such as nonlinear hyperbolic/parabolic partial differential equations (PDEs), Navier–Stokes equations, and population balance equations are employed as the basis for the design of high-performance feedback controllers used to regulate spatial temperature and concentration profiles in advanced materials processing applications, achieve wave suppression and drag reduction in fluid dynamic systems, and shape particle size distribution in particulate processes, respectively (see, for example, the special volumes [34, 33] and the books [28, 29] for representative results and references in these areas). On the other hand, for processes that involve coupling of macroscale phenomena with important phenomena at mesoscopic/microscopic length scales, multiscale systems coupling continuum-type distributed parameter systems with molecular dynamics (MD) or kinetic Monte Carlo (MC/kMC) simulations are employed because of their ability to describe phenomena that are inaccessible with continuum laws and equations.

An industrially important process where multiscale modeling is needed to adequately describe the coupling of macroscopic and microscopic phenomena is thin-film growth. Thin films of advanced materials are currently used in a very wide range of applications, e.g., microelectronic devices, optics, micro-electro-mechanical systems (MEMS), and biomedical products. Various deposition methods have been developed and widely used to prepare thin films such as physical vapor deposition (PVD) and chemical vapor deposition (CVD). However, the dependence of the thin-film properties, such as uniformity, composition, and microstructure, on the deposition conditions is a severe constraint on reproducing the thin film's performance. Thus, real-time feedback control of thin-film deposition, based on fundamental

P.D. Christofides et al., *Control and Optimization of Multiscale Process Systems*,
Control Engineering, DOI 10.1007/978-0-8176-4793-3_1,
© Birkhäuser Boston, a part of Springer Science+Business Media, LLC 2009

models, becomes increasingly important in order to meet the stringent requirements on the quality of thin films and reduce thin-film variability. While deposition uniformity and composition control can be accomplished on the basis of continuum-type distributed parameter models [see, for example, [28, 147] for results on rapid thermal processing (RTP) and [6, 119] on plasma-enhanced chemical vapor deposition (PECVD)], precise control of thin-film microstructure requires multiscale distributed models that predict how the film state (microscopic scale) is affected by changes in the controllable process parameters (macroscopic scale). In the remainder of this chapter, we discuss the implications of these problems in the context of chemical process control applications and review some of the relevant literature on this subject.

1.2 An Example of a Multiscale Process: Thin-Film Growth

Consider a conceptual thin-film growth process in a reactor with the split-inlet configuration shown in Fig. 1.1. The bulk of the reactor can be modeled using two-dimensional axisymmetric PDEs in cylindrical coordinates derived from continuum conservation principles. The surface of the growing film is modeled using kMC simulations. Figure 1.1 also shows the domains of definition of the two models. It should be noted that the microscopic domain is infinitesimally thin. Substrate temperature profiles can be manipulated using three circular heaters with heat being conducted in the in-between areas. Gaseous species A and B represent the precursors of a and b (components of compound semiconductor ab), respectively, and are assumed to undergo the gas-phase reactions in the bulk of the reactor and gas-surface reactions on the wafer surface, shown in Table 1.1. Reaction $G1$ represents the thermal decomposition of precursor A into A', which adsorbs on the substrate (reaction $S1$). The rate parameter for adsorption of A' (reaction $S1$) is assumed to follow that of an ideal gas, i.e. $k_a = s_0 \sqrt{(RT)/(2\pi M)}$, where s_0 is the sticking coefficient. The rate of

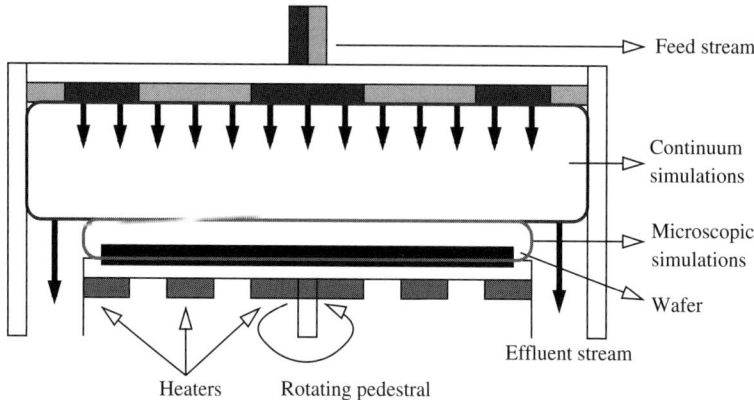

Fig. 1.1. Schematic of the reactor with a split-inlet configuration.

Table 1.1. Process reaction scheme.

Reaction
(G1) $A \rightarrow A' + C$
(S1) $A' \rightarrow a(s) + D$
(S2) $B \rightarrow b(s)$

adsorption of B (reaction $S2$) is assumed to be equal to $S1$ so that the stoichiometry of the film is preserved. In addition to adsorption, thin-film surface diffusion and desorption of adsorbed species are other significant processes that affect the morphology of the surface. The rate of desorption of surface species into the gas phase and the rate of surface diffusion are given by

$$k_d^n = k_{d0} e^{-\frac{E_{d0} + n\Delta E}{k_B T}}, \quad k_m^n = \frac{k_B T}{h} e^{-\frac{E + n\Delta E}{k_B T}}, \tag{1.1}$$

where h is Planck's constant, E and E_{d0} are the energy barriers for surface diffusion and desorption, respectively, ΔE is the interaction energy between two neighboring adsorbed species, and $n \in \{0, 1, 2, 3, 4\}$ is the number of nearest neighbors.

The macroscopic model for the gas phase of the reactor is given by the following conservation equations:

$$\nabla \cdot (\rho \mathbf{u}) = 0, \quad \nabla \cdot (\rho \mathbf{u}\, \mathbf{u}) - \nabla \cdot \mathbf{T} - \rho \mathbf{g} = 0,$$
$$\nabla \cdot (\rho \mathbf{u} T) = -\nabla \cdot \mathbf{q} - \sum_k h_k W_k \dot{\omega},$$
$$\nabla \cdot (\rho \mathbf{u} Y_k) = -\nabla \cdot \mathbf{j}_k + W_k \dot{\omega}_k, \quad k \in \{1, 2, 3, 4\},$$
$$\mathbf{j}_k = -D_k \rho \nabla Y_k - D_{T,k} \frac{\nabla T}{T}, \tag{1.2}$$

where ρ is the gas-phase density, $\mathbf{u}$ is the fluid velocity vector, $\mathbf{T}$ is the stress tensor, C_p is the specific heat capacity, T is the temperature, $\mathbf{q}$ is the heat flux due to conduction, and h_k, W_k, and Y_k are the partial specific enthalpy, molecular weight, and the mass fraction of gas species. $\dot{\omega}_k$ and $\mathbf{j}_k$ are the net production rate due to homogeneous reactions and mass flux, respectively, of species k. D_k and $D_{T,k}$ in the flux equation correspond to mass diffusion and thermal diffusion coefficients, respectively.

The flux boundary condition at the deposition surface is given by

$$\mathbf{j} = R_{ad} = k_a C_{A'}|_s - <k_d> f(C_{a.s}, T, w_{A'A'}), \tag{1.3}$$

where R_{ad} is the net rate of adsorption, T_s is the surface temperature, and $<k_d>$, $C_{A'}|_s$, and $C_{a.s}$ are the effective desorption rate, concentration of A' over the substrate, and *average* surface concentration of adsorbed $a(s)$, respectively. Function f describes the influence of lateral interactions on the desorption rate, which cannot be ascertained without knowledge of microscopic surface structure. Kinetic Monte Carlo models can be used to account for the surface microstructure and estimate the right-hand side of Eq. (1.3), which links the two levels of descriptions.

Kinetic Monte Carlo simulation approximates the solution of the following stochastic master equation [43] through Monte Carlo sampling:

$$\frac{\partial P(\sigma, t)}{\partial t} = \sum_{\sigma'} W(\sigma', \sigma) P(\sigma', t) - W(\sigma, \sigma') P(\sigma, t), \qquad (1.4)$$

where σ and σ' are surface configurations and $P(\sigma, t)$ is the probability that the surface configuration is in state σ at time t, and $W(\sigma, \sigma')$ is the probability per unit time of transition from σ to σ'. It is assumed that at any instant, only a single event (out of all possible events) occurs, according to its relative probability. After each event, time is incremented based on the total rate of all surface microprocesses.

A typical control problem for this process is to regulate the thin-film growth rate (macroscale objective) and the surface roughness (microscale objective) by manipulating the gas-phase composition and substrate temperature.

1.3 Background on Control and Optimization of Multiscale Process Systems

The objective of this section is to provide a review of results on control and optimization of multiscale systems coupling continuum-type distributed parameter systems with atomistic/particle simulations. We emphasize the progress recently made on the precise regulation of spatial temperature and concentration profiles in distributed process systems for which continuum laws are applicable and of material microstructure in advanced materials processing applications for which atomistic models are used. The review is not intended to be exhaustive; its objective is to provide the necessary background for the results of this book.

Research on the dynamics of distributed parameter systems has led to the discovery that the dominant dynamic behavior of many highly dissipative distributed process systems can be characterized by a small number of degrees of freedom, which has led to the introduction of rigorous mathematical concepts (e.g., inertial manifold (IM) [145]) to capture this type of behavior and to the development of advanced model-reduction techniques for deriving low-dimensional approximations that accurately reproduce the dynamics and solutions of various classes of infinite-dimensional systems. However, the explicit derivation of the inertial form requires the computation of the analytic form of the IM. Unfortunately, IMs have been proven to exist only for certain classes of PDEs (for example, Kuramoto–Sivashinsky equation and some diffusion-reaction equations [145]), and even then it is almost impossible to derive their analytic form. In order to overcome the problems associated with the existence and construction of IMs, the concept of approximate inertial manifold (AIM) has been introduced (see, for example, [47, 46, 148, 77]) and used for the derivation of ODE systems whose dynamic behavior approximates that of the inertial form.

More recently, significant developments in control nonlinear distributed parameter systems have been accomplished by bringing together concepts from nonlinear dynamics of infinite-dimensional systems and nonlinear control theory. Specifically, research has led to the development of a general and practical framework for the synthesis of nonlinear low-order feedback controllers used to regulate spatial temperature and concentration profiles in advanced materials processing applications, achieve wave suppression and drag reduction in fluid dynamic systems, and shape particle size distribution in particulate processes, respectively (see the books [28, 29] for results). The key idea is the development of a singular perturbation formulation of Galerkin's method that leads to a practical procedure for the construction of approximate inertial manifolds (AIMs) of highly dissipative, infinite-dimensional systems. The AIMs are used to derive accurate low-order approximations of the PDE systems that form the basis for the synthesis of nonlinear low-order output feedback controllers. Within the developed framework, the infinite-dimensional, closed-loop system stability, performance, and robustness properties have been precisely characterized in terms of the accuracy of the approximation of the low-dimensional models. Due to the low-dimensional structure of the controllers, the computation of the control action involves the solution of a small set of ordinary differential equations (ODEs); thus, the developed controllers can be readily implemented in real time with reasonable computing power.

However, many applications of industrial relevance require product quality specifications that are characterized by phenomena that evolve at both macroscopic and microscopic length scales. In general, continuum mathematical descriptions that apply to macroscopic phenomena are inadequate at microscopic length scales. Atomistic/particle simulation techniques such as molecular dynamics, kinetic Monte Carlo, and Lattice–Boltzmann (LB) simulations, on the other hand, are computationally too expensive to be employed for macroscopic process domains. This problem has been addressed through the development of hybrid continuum/atomistic multiscale models that augment the continuum macroscopic description of a process by embedding microscopic descriptions *only* at small subdomains of the process where submacroscopic resolution is required. Such models have been developed for fluid flows [120, 52, 96, 44], moving contact line [66], transient fluid flows with heat transfer [38], crack propagation [23, 114], chemical vapor deposition [155, 76, 156, 125, 90, 112], and biological systems [108, 161, 4], to name a few. In most of these multiscale frameworks, the inner (microscopic) and outer (macroscopic) models evolve concurrently and interact through exchange of particles such that mass, momentum, and energy remain conserved.

Fundamental mathematical modeling techniques have been developed to describe the microscopic features of surfaces formed by surface microprocesses, which include (1) kinetic Monte Carlo methods [57, 43, 133, 127] and (2) stochastic partial differential equations [39, 158, 36, 91]. The kinetic Monte Carlo simulation methods can be used to predict average properties of thin films (which are of interest from a control point of view; for example, surface roughness), by explicitly accounting for the microprocesses that directly shape thin-film microstructure. Stochastic PDEs contain the surface morphology information of thin films, and, thus, they may be used for the purpose of feedback controller design. For example, it has been experimentally verified that the Kardar–Parisi–Zhang (KPZ) equation [82] can describe the evolution of the surface morphology of gallium arsenide (GaAs) thin films, which is consistent with the surface measured by atomic force microscopy (AFM) [13, 79]. Furthermore, based on the fact that kinetic Monte Carlo simulations provide realizations of a stochastic process that are consistent with the master equation that describes the evolution of the probability distribution of the system being at a certain microconfiguration, a method to construct reduced-order approximations of the master equation was reported in [51]. Recently, a method was also developed to identify an empirical input–output model for a copper electrodeposition process using simulation data from a coupled kMC and finite-difference simulation code and to perform controller design using the identified model [130].

In the context of control of processes described by microscopic simulations, a so-called coarse time-stepper approach [146, 53, 106, 54] has been utilized to design linear discrete-time controllers for lumped [137, 134] and distributed processes [11, 136] described by microscopic simulations. This approach circumvents the derivation of a closed-form macroscopic model for the process by identifying the essential coarse-scale system behavior and is effective in controlling macroscopic variables that are low statistical moments of the microscopic distributions (e.g., surface coverage, which is the zeroth moment of species distribution on a lattice). However, to control higher statistical moments of the microscopic distributions, such as the surface roughness (the second moment of height distribution on a lattice) or even the microscopic configuration (such as the surface morphology), deterministic models may not be sufficient. This is because the effect of the stochastic nature of the microscopic processes becomes very significant in these cases and must be addressed in both the model construction and the controller design.

The problem of optimal operation of multiscale processes is gaining considerable significance due to increasingly more stringent performance specifications. This has led to the incorporation of detailed process models into process optimization frameworks. Nonlinear model-reduction techniques for transport-reaction processes operating at steady state [17, 18] and unsteady state [9] were recently employed to formulate approximate low-order optimization problems. In these investigations spatial discretization was carried out using the method of weighted residuals with empirical eigenfunctions as basis functions. These eigenfunctions were generated by the application of Karhunen–Loève expansion (KLE, also known as proper orthogonal decomposition, principal component analysis, and method of empirical eigenfunctions [138, 139]) on an ensemble of solution data of the PDEs for the span of process parameters. The motivation behind this approach was the presence of a finite number of dominant spatial patterns (eigenmodes) in the solution of highly dissipative PDEs that govern its long-time dynamics, while the remaining infinite-dimensional (stable) fast modes relax to these finite-dimensional slow dynamics [28, 145]. The nonlinear programs (NLPs) thus formed are significantly lower in size and can be solved faster than those that result from discretization of the distributed parameter system with finite differences/elements. The principal reason that allows model reduction is that the spatiotemporal behavior of the given PDE system is accounted for in the shape of the empirical eigenfunctions. An additional reduction in the size of the NLP can be made if temporal discretization is performed *only* for the vector of control variables and ODE equality constraints are directly integrated in time (the so-called control vector parametrization (CVP) scheme [19, 21, 41, 153, 131]). The advantage of the control vector parameterization-based scheme is that optimization is performed for the reduced set of discretized decision variables rather than for the complete set of discretized variables. Incorporation of second-order derivative information into the CVP framework for improved efficiency has also been addressed in [14, 15]. Hence, for distributed processes, the combination of spatial discretization using Galerkin/KLE and temporal discretization using CVP offers an attractive strategy for the efficient solution of the corresponding optimal control problems.

Process optimization with microscopic simulations presents a distinct set of challenges. Unlike continuum models for macroscopic phenomena, microscopic models are unavailable in closed form. This implies that the corresponding optimal control problem is constrained by dynamic equalities whose explicit form is unavailable. Moreover, since microscopic models are stochastic in nature, traditional gradient-based algorithms have limited applicability. The standard approach for the solution of such problems is to compute the objective functional as a "black box" and employ direct search algorithms such as Hooke–Jeeves, Nelder–Mead, pattern search [149, 94, 95], etc. to compute the optimal control trajectory [136, 11]. An alternative methodology for the global optimization of nonlinear programs constrained by "nonfactorable" constraints (constraints defined by a computational model for which no explicit analytical representation is available) was proposed in [113]. However, the above approaches are inefficient if the computation of the cost functional (or the black-box simulation) is expensive, which is usually the case with microscopic simulations. A number of approaches are specially designed for problems where computation of the objective function is expensive [22, 78], but most of them have been employed for noise-free systems, which limits their applicability to the case of optimization of processes described by microscopic simulations.

1.4 Objectives and Organization of the Book

Motivated by the industrial significance of multiscale process systems and the lack of general control and optimization methods for such systems, the broad objectives of this book are as follows:

- To present a framework for the design of real-time control systems that systematically integrates fundamental process models, model-reduction techniques, feedback control laws, and real-time surface measurement techniques to regulate material microstructure in multiscale processes.
- To present a methodology for the efficient solution of optimization problems when the cost functional and/or equality constraints span multiple length scales and necessitate multiscale process models.
- To provide fundamental understanding and insight into the nature of the control and optimization problems for process systems characterized by coupled macroscopic and microscopic phenomena.

- To illustrate the application of the proposed control and optimization methods to material preparation processes of practical interest and document their effectiveness and advantages with respect to existing control and optimization methods.

The rest of the book is organized as follows. Chapter 2 reviews the challenges and modeling approaches for multiscale processes and provides the necessary background for presenting our methods for control and optimization of multiscale process systems. A detailed multiscale model of a thin-film growth process in a stagnation point geometry is presented. Specifically, a set of PDEs is used to model the gas-phase dynamics and the kinetic Monte Carlo model is used to model the thin-film surface microstructure. The theoretical foundation of the kinetic Monte Carlo simulation is reviewed. It is demonstrated that by assuming the surface microprocesses are Poisson processes, both the master equation and the kinetic Monte Carlo simulation algorithm can be derived. Finally, a tutorial is given to show the details of the algorithm for kinetic Monte Carlo simulation of a thin-film growth process.

Chapter 3 addresses the real-time feedback control of thin-film growth using kinetic Monte Carlo models. A real-time estimator is first constructed that allows estimates of the surface roughness and growth rate to be estimated at a time scale comparable to the real-time evolution of the process. The real-time estimates enable the design of feedback controllers to regulate the thin-film growth process. Methods for the design of single-input–single-output (SISO), multiple-input–multiple-output (MIMO), and predictive controllers using kMC models are then presented. The applications of the developed real-time feedback control methods are illustrated through two thin-film growth processes.

Stochastic partial differential equations (PDEs), as closed-form microscopic models, arise naturally in the modeling of the evolution of the surface height profile of ultra-thin films in a variety of material preparation processes and can be used as a basis for model-based controller synthesis. Chapter 4 focuses on the methods for the construction of linear and nonlinear stochastic PDEs. The construction of linear stochastic PDE models for thin-film deposition processes involves the derivation of the analytical solution for the statistical moments of the state of the process and the generation of surface snapshots for different instants during process evolution using kMC simulations. A linear stochastic PDE model is determined by least-squares-fitting model parameters to match the kMC simulation results. Nonlinear stochastic PDE models are constructed by combining a priori knowledge on the model structure and a novel model parameter estimation procedure. A deterministic finite-dimensional ODE system is first derived for the evolution of the state covariance matrix for the dominant modes of the stochastic PDE. Model parameters are subsequently estimated by using surface snapshots generated using a kMC simulation and solving an overdetermined least-squares minimization problem. The effectiveness of the methods is demonstrated through applications to two thin-film deposition processes and an ion-sputtering process.

In Chapter 5, we present methods for model-based controller design based on the stochastic PDEs to control the thin-film surface roughness. A method for multivariable model predictive control using linear stochastic PDEs is first presented.

The method results in the design of a computationally efficient multivariable predictive control algorithm that is successfully applied to the kMC models of thin-film deposition processes taking place on both 1D and 2D lattices to regulate the thin-film thickness and surface roughness at desired levels. When spatially distributed sensing and actuation are available, the surface roughness can be regulated using the covariance control technique. A method for linear covariance controller design is first developed and is demonstrated through application to the kMC model of a 1D thin-film growth process whose surface height evolution can be described by the Edwards–Wilkinson equation. Motivated by the fact that nonlinearities exist in many material preparation processes in which the surface evolution can be modeled by nonlinear stochastic PDEs, we also propose a method for nonlinear covariance controller design based on nonlinear stochastic PDEs. The stochastic KSE is used to evaluate the developed method. The proposed nonlinear controller is applied to a high-order approximation of the stochastic KSE and the kMC model of an ion-sputtering process.

Chapter 6 addresses the issue of efficient solution of optimization problems when the cost functional and/or equality constraints span multiple length scales and necessitate multiscale process models. We consider a conceptual thin-film epitaxy process and optimize the process operation for two simultaneous objectives that span multiple length scales: (1) to maximize the thickness uniformity of the deposited film (macroscopic objective) and (2) to minimize the surface roughness of the deposited film (microscopic objective) across the wafer surface at the end of the process cycle. A multiscale process model is formulated linking continuum conservation laws for reactor-scale phenomena and kMC simulations for the microscopic film-surface processes. The computational intensity of the multiscale process model prohibits its direct incorporation into process optimization. To address this issue, order-reduction techniques for dissipative PDEs are linked with the adaptive tabulation scheme for the solution data from the microscopic model to derive a computationally efficient multiscale model that forms the equality constraints of the optimization problem. The process is modeled mathematically using continuum conservation laws and the microscopic film-surface processes are modeled using kMC simulations. Initially, we calculate optimal substrate temperature profiles for a steady-state process operation such that the grown thin films have a high degree of spatial uniformity and, simultaneously, low surface roughness. Subsequently, we improve the process operation by computing time-varying substrate temperature radial profiles and inlet concentration profiles of the precursors to meet the optimization objectives.

Chapter 7 extends the methodology outlined in the previous chapter to efficient solution of dynamic optimization problems for multiscale processes coupling continuum and discrete descriptions. The approach relies on the reduction of the continuum system using Karhunen–Loève expansion (KLE) and the discrete system using in situ adaptive tabulation (ISAT). The reduced systems are linked together to formulate a computationally efficient multiscale model. Consequently, standard search algorithms are employed for the solution of the optimization problem. The approach is demonstrated on two numerical examples describing catalytic oxidation of CO to CO_2. In the first case, lateral interactions between adsorbed species are neglected and the mobility of adsorbed species is assumed to be infinitely fast, so that the system can be approximated as a well-mixed system. In the latter case, nearest-neighbor interactions between adsorbed CO molecules are introduced along with the finite mobility of the adsorbed species. The microscopic system is linked to a macroscopic system describing diffusion of CO and O_2 on the catalyst surface. In both cases, optimal inlet concentration profiles are computed to guide the microscopic system from one stable stationary state to another stable stationary state.

2

Multiscale Process Modeling and Simulation

2.1 Overview of Multiscale Modeling

A large number of the processes relevant to the chemical process industry necessitate the consideration of transport phenomena (fluid flow, heat and mass transfer), often coupled with chemical reactions. Examples range from reactive distillation in petroleum processing to plasma-enhanced chemical vapor deposition, etching, and metalorganic vapor-phase epitaxy in semiconductor manufacturing. Mathematical descriptions of these transport-reaction processes can be derived from dynamic conservation equations and usually are comprised of highly dissipative partial differential equations (PDEs). In addition, efforts have also gained momentum toward detailed modeling of processes at microscopic length scales. To this end, atomistic/particle simulation techniques such as molecular dynamics (MD), kinetic Monte Carlo (KMC), Lattice–Boltzmann (LB), etc. have been utilized in place of traditional continuum or mean-field approaches. The advantage of using atomistic models is their capability to describe phenomena whose characteristic length and time scales are much smaller than those for which the continuum approximation holds. Examples where microscopic simulations have been utilized include modeling of homogeneous reacting systems [57, 59, 5], biological systems [110, 126, 140], microstructure evolution during thin-film growth [133, 61, 109, 60], crack propagation [143], and fluid flow [96, 16], to name a few.

An industrially important process where multiscale modeling is needed to adequately describe the coupling of macroscopic and microscopic phenomena is thin-film growth. Thin films of advanced materials are currently used in a very wide range of applications, e.g., microelectronic devices, optics, micro-electro-mechanical systems (MEMS), and biomedical products. Various deposition methods have been developed and widely used to prepare thin films such as physical vapor deposition (PVD) and chemical vapor deposition (CVD). In the remainder of this chapter, we discuss a detailed multiscale model of a thin-film growth process.

P.D. Christofides et al., *Control and Optimization of Multiscale Process Systems*, Control Engineering, DOI 10.1007/978-0-8176-4793-3_2,

2.2 Thin-Film Growth Process

In this section, in order to discuss an example of multiscale modeling and provide the necessary background for presenting our methods for control and optimization of process systems using multiscale models, we consider the growth of a thin film from a fluid in a vertical, stagnation-flow geometry. The process is shown in Fig. 2.1. In this geometry, the inlet fluid flow forms a uniform boundary layer adjacent to the surface of the substrate and precursor atoms diffuse through the boundary layer and deposit a thin film [49]. Upon arrival at the surface, the precursor atoms are adsorbed onto the surface. Subsequently, adsorbed atoms may desorb to the gas phase or migrate on the surface.

From a modeling point of view, the major challenge is the integration of the wide range of length and time scales that the process encompasses [155]. Specifically, in the gas phase, the processes of heat/mass transport can be adequately modeled under the hypothesis of continuum, thereby leading to PDE models for chamber temperature and species concentration. However, when the microstructure of the surface is studied, microscopic events such as atom adsorption, desorption, and migration have to be considered, and the length scale of interest reduces dramatically to the order of that of several atoms. Under such a small length scale, the continuum hypothesis is no longer valid and deterministic PDEs cannot be used to describe the microscopic phenomena. Microscopic simulation techniques should be employed to describe the evolution of the surface microstructure.

Although different modeling approaches are needed to model the macroscopic and microscopic phenomena of the process, there are strong interactions between the macro- and microscale phenomena. For example, the concentration of the precursor in the inlet gas governs the rate of adsorption of atoms on the surface, which, in turn, influences the surface roughness. On the other hand, the density of the adatoms on the surface affects the rate of desorption of atoms from the surface to the gas phase, which, in turn, influences the gas-phase concentration of the precursor. A multiscale model [155] is employed in this work to capture the evolution of both macroscopic and microscopic phenomena of the thin-film growth process as well as their interactions. A set of PDEs derived from the mass, momentum, and energy balances is

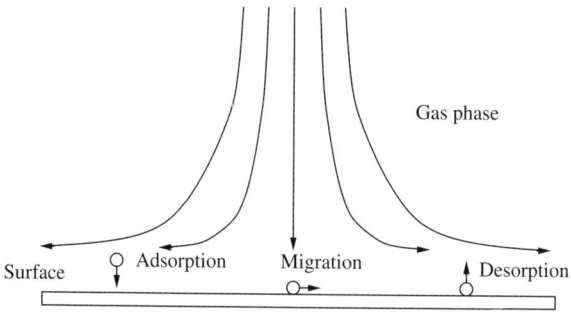

Fig. 2.1. Illustration of the thin-film growth process.

used to describe the gas-phase dynamics. Kinetic Monte Carlo (MC) simulation is employed to capture the evolution of the surface microstructure. Furthermore, the parameters of MC simulation such as the temperature and precursor concentration are provided by the solution of PDE, and the results from the kinetic MC simulation are used to determine the boundary conditions of the PDEs of the macroscopic model. In the remainder of this section, we describe the model for the gas phase and the surface microstructure for the thin-film growth process of Fig. 2.1.

2.3 Gas-Phase Model

Under the assumption of axisymmetric flow, the gas phase can be modeled through continuum-type momentum, energy, and mass balances as follows [90]:

$$\frac{\partial}{\partial \tau}\left(\frac{\partial f}{\partial \eta}\right) = \frac{\partial^3 f}{\partial \eta^3} + f\frac{\partial^2 f}{\partial \eta^2} + \frac{1}{2}\left[\frac{\rho_b}{\rho} - \left(\frac{\partial f}{\partial \eta}\right)^2\right], \tag{2.1}$$

$$\frac{\partial T}{\partial \tau} = \frac{1}{P_r}\frac{\partial^2 T}{\partial \eta^2} + f\frac{\partial T}{\partial \eta}, \tag{2.2}$$

$$\frac{\partial y_i}{\partial \tau} = \frac{1}{Sc_j}\frac{\partial^2 y_i}{\partial \eta^2} + f\frac{\partial y_i}{\partial \eta}. \tag{2.3}$$

The following boundary conditions are used for $\eta \to \infty$:

$$T = T_{\text{bulk}}, \qquad \frac{\partial f}{\partial \eta} = 1, \qquad y_j = y_{jb}, \qquad j = 1, \dots, N_g, \tag{2.4}$$

and for $\eta \to 0$ (surface):

$$T = T_{\text{surface}}, \qquad f = 0, \qquad \frac{\partial f}{\partial \eta} = 0,$$
$$\frac{\partial y_j}{\partial \eta} = 0, \text{ for } j \neq \text{growing}, \tag{2.5}$$
$$\frac{\partial y_{\text{growing}}}{\partial \eta} = \frac{Sc_{\text{growing}}(r_a - r_d)}{\sqrt{2a\mu_b\rho_b}},$$

where f is the dimensionless stream function, r_a and r_d are the rates of adsorption and desorption, respectively, η is the dimensionless distance to the surface, ρ is the density of the mixture, P_r is the Prandtl number, y_j and Sc_j are the mole fraction and Schmidt number of species j, respectively, μ_b and ρ_b are the viscosity and the density of the bulk, respectively, a is the hydrodynamic strain rate, and $\tau = 2at$ is the dimensionless time.

Although the macroscopic model describes the evolution of the precursor concentration and temperature (which influence the configuration of the growing surface), no direct information about the surface microstructure is available from the macroscopic model. Furthermore, the boundary conditions for the mass-transfer equation of the growing species depend on the rate of adsorption and desorption. Therefore, a microscopic model is necessary to model the surface microstructure and to determine the boundary conditions of the mass-transfer equation.

2.4 Surface Microstructure Model

The thin-film growth of Fig. 2.1 includes three processes: the adsorption of atoms from the gas phase to the surface, the desorption of atoms from the surface to the gas phase, and the migration of atoms on the surface. In this study, we consider multilayer growth and assume that all the surface sites are available for adsorption at all times, therefore, the adsorption rate is treated as site-independent. For an ideal gas, the adsorption rate is given by the kinetic theory [90]

$$r_a = \frac{s_0 P}{2a\sqrt{2\pi mkT} C_{\text{tot}}}, \tag{2.6}$$

where s_0 is the sticking coefficient, k is the Boltzmann constant, P is the partial pressure of the precursor, C_{tot} is the concentration of sites on the surface, m is the molecular weight of the precursor, T is the gas-phase temperature above the surface, and a is the strain rate. The rate of desorption of an atom depends on the atom's local microenvironment (i.e., interactions with nearest neighbors) and the local activation energy. Under the consideration of only first nearest-neighbor interactions, the desorption rate of an atom from the surface with n first nearest neighbors is

$$r_d(n) = \frac{\nu_0}{2a} \exp\left(-\frac{nE}{kT}\right), \tag{2.7}$$

where E is the energy associated with a single bond on the surface and ν_0 is the frequency of events, which is determined by the following expression:

$$\nu_0 = k_{d0} \exp\left(-\frac{E_d}{kT}\right), \tag{2.8}$$

where k_{d0} is an event-frequency constant and E_d is the energy associated with desorption. Finally, the surface migration is modeled as desorption followed by re-adsorption [60], and the migration rate is given by

$$r_m(n) = \frac{\nu_0 A}{2a} \exp\left(-\frac{nE}{kT}\right), \tag{2.9}$$

where A is associated with the energy difference that an atom on a flat surface has to overcome in jumping from one lattice site to an adjacent one and A is given as

$$A - \exp\left(\frac{E_d - E_m}{kT}\right), \tag{2.10}$$

where E_m is the energy associated with migration.

2.4.1 Poisson Processes and Master Equation

The formation of the thin film by adsorption, migration, and desorption is a stochastic process because (1) the exact time and location of the occurrence of one specific surface microprocess (adsorption, migration, or desorption) are unknown, and (2) the probability with which each surface microprocess may occur is only available. Therefore, the surface evolution model should be established based on probability theory.

Specifically, we treat the surface microprocesses as Poisson processes, which means that the following assumptions are made [111, 42, 57, 43]:

Assumption 2.1. *The probability that k events occur in the time interval $(t, t + T)$ is independent of t.*

Assumption 2.2. *The probability that k events occur in the time interval $(t, t+T)$ is independent of the number of events occurring in any nonoverlapping time interval.*

Assumption 2.3. *The probability that an event occurs in an infinitesimal time interval $[t, t + dt)$ is equal to $W \cdot dt$ (where W is the mean count rate of the event), and the probability of more than one event occurring in an infinitesimal time interval is negligible.*

Based on these three assumptions, the time evolution of probabilities that the surface is in one specific configuration can be derived. The configuration of a surface is characterized as the height of each surface atom at each surface site. If $P(\alpha, t)$ represents the probability that the system is in configuration α at time t, based on Assumptions 2.2 and 2.3, we have the following equation for $P(\alpha, t + dt)$:

$$P(\alpha, t + dt) = P(\alpha, t)P_{0\alpha} + \sum_\beta P(\beta, t)P_{1\beta}, \qquad (2.11)$$

where $P_{0\alpha}$ is the probability that no event occurs in the time interval $(t, t+dt)$ given that the surface is in configuration α at t, $P(\beta, t)$ is the probability that the surface is in configuration β at t, and $P_{1\beta}$ is the probability that one event occurs in the time interval $(t, t+dt)$ given that the surface is in configuration β at t, and the occurrence of this event results in a transition from configuration β to configuration α.

$P_{0\alpha}$ and $P_{1\beta}$ have the following expression (a detailed proof can be found in [58]). Specifically,

$$P_{0\alpha} = 1 - \sum_\beta W_{\beta\alpha}dt, \qquad (2.12)$$

where $W_{\beta\alpha}dt$ is the probability that an event occurs in the time interval $[t, t + dt)$, which results in a transition from configuration α to a configuration β; therefore, $\sum_\beta W_{\beta\alpha}dt$ is the probability that any one event occurs in the time interval $(t, t+dt)$ provided that the surface configuration is α at t. Moreover,

$$P_{1\beta} = W_{\alpha\beta}dt, \qquad (2.13)$$

where $W_{\alpha\beta}dt$ is the probability that an event occurs in the time interval $[t, t + dt)$ and the occurrence of this event results in a transition from configuration β to configuration α.

By substituting Eqs. (2.12) and (2.13) into Eq. (2.11) and taking the limit $dt \to 0$, we obtain a differential equation describing the time evolution of the probability that the surface is in configuration α:

$$\frac{dP(\alpha, t)}{dt} = \sum_{\beta} P(\beta, t) W_{\alpha\beta} - \sum_{\beta} P(\alpha, t) W_{\beta\alpha}. \tag{2.14}$$

Eq. (2.14) is the so-called 'master equation' (ME) for a stochastic process. The ME has a simple, linear structure; however, it is difficult to write the explicit form of Eq. (2.14) for any realistic system because the number of the possible states is extremely large for most systems of a realistic size. For example, for a system with 10×10 sites and a maximum height of 1, the number of configurations is $2^{100} \approx 10^{30}$. This makes the direct solution of Eq. (2.14), for any system of meaningful size, using numerical methods for integration of ordinary differential equations (e.g., Runge–Kutta) impossible.

2.4.2 Theoretical Foundation of Kinetic Monte Carlo Simulation

Monte Carlo techniques provide a way to obtain unbiased realizations of a stochastic process, which is consistent with the ME. The consistency of the Monte Carlo simulation to the ME is based on the fact that in a Monte Carlo simulation, the time sequence of Monte Carlo events is constructed following a probability density function that is derived based on the same assumptions (Assumptions 2.1–2.3) as those used in the derivation of the master equation [57].

A Monte Carlo event is characterized by both the type of the event and the site in which the event is executed. We use $e(x; i, j)$ to represent a Monte Carlo event of type x executed on the site (i, j), where $x \in \{a, m, d\}$, where $x = a$ corresponds to an adsorption event, $x = m$ corresponds to a migration event, and $x = d$ corresponds to a desorption event, $1 \leq i, j \leq N$, and $N \times N$ is the size of the lattice.

The sequence of Monte Carlo events can be constructed based on the probability density function, $F(\tau, e)$, defined as follows:

Definition 2.1. $F(\tau, e)d\tau$ *is the probability at time t that event e will occur in the infinitesimal time interval* $(t + \tau, t + \tau + d\tau)$.

We now compute the expression of $F(\tau, e)$ based on Assumptions 2.1 and 2.2:

$$F(\tau, e)d\tau = P_{0\tau}P_e, \tag{2.15}$$

where $P_{0\tau}$ is the probability that no event occurs in $[t, t+\tau)$, and P_e is the probability that event e occurs in the time interval $(t + \tau, t + \tau + d\tau)$. P_e can be determined by using Assumption 2.3 as follows:

$$P_e = W_e d\tau. \tag{2.16}$$

$P_{0\tau}$ can be calculated by sampling the duration $(t, t + \tau)$ into M identical time intervals $\delta\tau = \tau/M$. When $M \to \infty$, $\delta\tau$ is small enough so that each time interval of size $\delta\tau$ contains one event at most. Based on Assumption 2.3, the probability that one event e occurs in $\delta\tau$ is $W_e\delta\tau$ and, based on Assumption 2.2, the probability that any one event occurs in $\delta\tau$ is $\sum_e W_e\delta\tau$. Therefore, the probability that no events will occur in $\delta\tau$ is

$$P_{0\delta\tau} = 1 - \sum_e W_e\delta\tau, \qquad (2.17)$$

where $P_{0\delta\tau}$ is the probability that no event occurs in one $\delta\tau$ interval and $W_e\delta\tau$ is the probability that one event e will occur in the $\delta\tau$ interval.

Eq. (2.17) can be applied to all the $\delta\tau$ time intervals in the duration $(t, t + \tau)$. Therefore, the probability that no events will occur in the duration τ is

$$P_{0\tau} = \lim_{N\to\infty} P_{0\delta\tau}^N = \lim_{N\to\infty} \left(1 - \frac{\sum_e W_e\tau}{N}\right)^N = \exp\left(-\sum_e W_e\tau\right). \qquad (2.18)$$

By substituting Eqs. (2.18) and (2.16) into (2.15) and using $W_{\text{tot}} = \sum_e W_e$, the probability density function, $F(\tau, e)$, is as follows:

$$F(\tau, e) = W_e \exp(-W_{\text{tot}}\tau). \qquad (2.19)$$

Monte Carlo simulation constructs a sequence of events following the probability density function shown in Eq. (2.19). Note that Eq. (2.19) is based on the same assumptions as the master equation [Eq. (2.14)]; therefore, the kinetic Monte Carlo simulation is able to provide an unbiased realization of a stochastic process that is consistent with that described by the master equation. Many kinetic Monte Carlo algorithms are available to simulate a stochastic dynamic process. In the remainder of this section, we discuss in detail the theoretical foundation and steps in the so-called "direct" method developed by [57] and of the algorithm used in the calculations. The "direct" method is based on the fact that the two-variable probability density function, Eq. (2.19), can be written as the product of two one-variable probability functions:

$$F(\tau, e)d\tau = F_1(\tau)d\tau \cdot P(e|\tau), \qquad (2.20)$$

where $F(\tau, e)d\tau$ is the probability that event e will occur in the time interval $(t + \tau, t + \tau + d\tau)$, $F_1(\tau)d\tau$ is the probability that an event will occur in the time interval $(t + \tau, t + \tau + d\tau)$, and $P(e|\tau)$ is the probability that the next event will be event e, given that the next event will occur in $(t + \tau, t + \tau + d\tau)$.

Based on the addition theorem [111], $F_1(\tau)d\tau$ is the sum of $F(\tau, e)d\tau$ over all events:

$$F_1(\tau)d\tau = \sum_e F(\tau, e)d\tau. \qquad (2.21)$$

$P(e|\tau)$ can be obtained by substituting Eq. (2.21) into Eq. (2.20):

$$P(e|\tau) = \frac{F(\tau, e)}{\sum_e F(\tau, e)}, \qquad (2.22)$$

By substituting Eq. (2.19) into Eqs. (2.21) and (2.22), we obtain

$$F_1(\tau) = W_{\text{tot}} \exp\left(-W_{\text{tot}}\tau\right), \qquad (2.23)$$

$$P(e|\tau) = \frac{W_e}{W_{\text{tot}}}, \qquad (2.24)$$

In the Monte Carlo simulation, Eq. (2.23) is used to determine the lifetime of a Monte Carlo event and Eq. (2.24) is used to determine the Monte Carlo event to be executed. To execute a Monte Carlo simulation, a pseudo-random number generator is used that generates random numbers following the uniform distribution in the interval $(0, 1)$. It has been proven by [57] that if a random number, ξ, follows the uniform distribution in the unit interval, then the lifetime of a Monte Carlo event, τ, can be computed by

$$\tau = -\frac{\ln \xi}{W_{\text{tot}}}. \qquad (2.25)$$

To demonstrate that the τ obtained by using Eq. (2.25) follows the probability density function in Eq. (2.23), we first compute the probability that $\tau < T$, $P(\tau < T)$, using Eq. (2.25). Specifically, we have

$$P(\tau < T) = P\left(-\frac{\ln \xi}{W_{\text{tot}}} < T\right) = P\left(\exp(-W_{\text{tot}}T) < \xi < 1\right). \qquad (2.26)$$

Because ξ follows the uniform distribution in the interval $(0, 1)$, from Eq. (2.26) we have that

$$P(\tau < T) = P(\exp(-W_{\text{tot}}T) < \xi < 1) = 1 - \exp(-W_{\text{tot}}T), \qquad (2.27)$$

whose corresponding probability density function, $F_1'(\tau)$, is

$$F_1'(\tau) = \frac{dP(\tau < T)}{dT} = W_{\text{tot}} \exp(-W_{\text{tot}}T), \qquad (2.28)$$

which is the probability density function in Eq. (2.23). Therefore, the lifetime of each Monte Carlo event, τ, calculated using Eq. (2.25) follows the probability density function in Eq. (2.23), which is consistent with the master equation.

2.4.3 Kinetic Monte Carlo Simulation of Thin-Film Growth

In this section, we show the details of kinetic Monte Carlo simulation of thin-film growth using the surface microstructure model presented above. We will demonstrate the initialization, Monte Carlo event selection, execution of adsorption, desorption, and migration events, computation of real time, and updating of the classes. In the simulation, we use the (001) surface of a simple cubic lattice with periodic boundary conditions. The size of the simulation lattice is 10×10. The adsorption rate is site-independent and is set to be $100/\text{s}$. The substrate temperature is fixed at $800\,\text{K}$. Both the bond energy and desorption energy are $0.74\,\text{eV}$ and the migration energy is $0.44\,\text{eV}$. The strain rate is $5\,\text{s}^{-1}$. The desorption rate is $4.06 \times \exp(-10n)/\text{s}$, and the migration rate is $233.1 \times \exp(-10n)/\text{s}$, where n is the number of nearest neighbors. Note that the three rates are determined in a way that is consistent with Eqs. (2.6)–(2.9).

Initializations

To start the simulation, the initial surface configuration should be specified. Although there is no limitation on the initial surface configuration, in many applications, a perfect surface is assumed to start the simulation [97, 100, 155]. In computer memory, the surface configuration can be stored in the form of a 2D array S, whose size is equal to the size of the simulation lattice. Each element indicates the number of atoms on the corresponding surface site of the substrate. In this simulation, a 10×10 zero matrix can be used to represent the initial perfect surface of the 10×10 lattice (flat surface with no atoms). A surface site is determined by its coordinates, (i, j), where i and j are integers satisfying $1 \le i \le N$ and $1 \le j \le N$ and N is the size of the lattice. When a 10×10 lattice is used in this simulation, $N = 10$. A (001) surface of a simple cubic lattice is used in this simulation. The initial surface configuration and the coordinates are shown in Fig. 2.2. Note that the same coordinate system will be applied to all simulations in this section and will not be marked in figures of surface configuration in the remainder of this section.

In the kMC simulation, all surface atoms need to be grouped according to the number of their nearest neighbors. On the (001) surface of a simple cubic lattice, a surface atom could have 0, 1, 2, 3, or 4 nearest neighbors. When the initial surface is a perfect surface, there are no surface atoms. The groups of surface atoms with 0, 1, 2, 3, and 4 nearest neighbors are initialized as empty sets.

Finally, we need to initialize the total rates of adsorption, desorption, and migration. The adsorption rate is site-independent and is set to be $100/\text{s}$ in this study. The initial total rate of adsorption is $1 \times 10^4/\text{s}$. Since there are no surface atoms on the

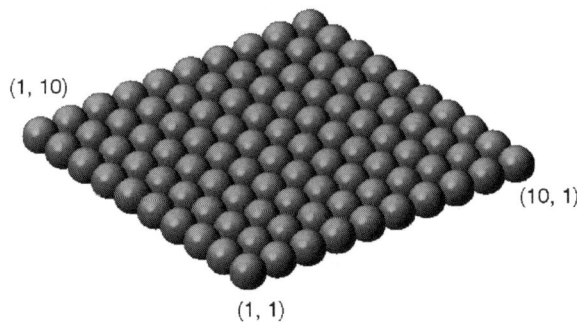

(1, 10)

(10, 1)

(1, 1)

Fig. 2.2. Initial surface configuration of the kMC simulation of the thin-film growth.

initial perfect surface, the initial values for the total rates of desorption and migration are both zero [see also Eqs. (2.29) – (2.31) below].

Monte Carlo Event Selection

The Monte Carlo algorithm picks an event to be executed based on the probability shown in Eq. (2.24). The probability of adsorption in an infinitesimal time interval $\delta\tau$ is site-independent, and the probabilities of desorption and migration are only dependent on the number of immediate side neighbors. To select a Monte Carlo event, the surface atoms are grouped into five classes based on the number of side neighbors (e.g., surface atoms have 0, 1, 2, 3, and 4 side neighbors); in each class, the atoms have the same desorption and migration probabilities (adsorption probability is site-independent). The total rate of adsorption, W_a, is computed as follows:

$$W_a = N^2 r_a, \tag{2.29}$$

where N is the size of the lattice and r_a is the adsorption rate on each surface site defined in Eq. (2.6).

The total rate of desorption, W_d, and the total rate of migration, W_m, are given by

$$W_d = \sum_{i=0}^{4} W_{d_i}, \qquad W_{d_i} = M_i \cdot r_d(i) \tag{2.30}$$

and

$$W_m = \sum_{i=0}^{4} W_{m_i}, \qquad W_{m_i} = M_i \cdot r_m(i), \tag{2.31}$$

where M_i is the number of surface atoms that have i side neighbors, which is equal to the number of atoms in each of the five classes, $r_d(i)$, defined in Eq. (2.7), is the desorption rate of an atom from the surface with i first nearest neighbors, and $r_m(i)$, defined in Eq. (2.9), is the migration rate of an atom from the surface with i first nearest neighbors.

To select a Monte Carlo event based on the rates, a random number following the uniform distribution in the unit interval, ζ, is generated. If $0 < \zeta < W_a/(W_a+W_d+W_m)$, the event is adsorption; if $W_a/(W_a + W_d + W_m) < \zeta < (W_a + W_d)/(W_a + W_d + W_m)$, the event is desorption; if $(W_a + W_d)/(W_a + W_d + W_m) < \zeta < 1$, the event is migration. Since the initial values for W_d and W_m are zero in this simulation, the first event must be an adsorption event.

Execution of a Monte Carlo Event: Adsorption

After the Monte Carlo event is selected, it needs to be executed. If the event is adsorption, the site on which the adsorption event will occur needs to be picked and the number of atoms on the selected site is increased by one. Since the adsorption rate is site-independent, the site for an adsorption event should be randomly picked

Fig. 2.3. The surface configuration after the execution of an adsorption event.

from all sites in the entire lattice. To randomly pick a site for an adsorption event, two random numbers uniformly distributed in the unit interval, $\zeta_{e,i}$ and $\zeta_{e,j}$, are first generated. The site is determined as $i = \text{ceil}(\zeta_{e,i} \times N)$ and $j = \text{ceil}(\zeta_{e,j} \times N)$, where the function $\text{ceil}(\cdot)$ rounds the element to the nearest integer toward infinity. In this simulation, we get $\zeta_{e,i} = 0.6068$ and $\zeta_{e,j} = 0.4860$. The site is, therefore, picked as $(7, 5)$. An atom is added to the site $(7, 5)$ as a result of the adsorption event. In computer memory, the 2D array S, which stores the surface configuration, is updated as $S(i, j) = S(i, j) + 1$ by assuming that its index begins with 1. The surface configuration after the execution of the adsorption event is shown in Fig. 2.3.

Remark 2.1. In above discussion, we assume that the array index begins with 1, which is the case for programming languages such as MATLAB and Fortran. Note that in some other programming languages, such as Java, C, and C++, array indices begin with 0. In that case, the 2D array S, which stores the surface configuration, should be updated as $S(i - 1, j - 1) = S(i - 1, j - 1) + 1$. In the remainder of this discussion, we consider the case where the array index begins with 1.

Compute the Real Time

Upon an executed event, a real-time increment Δt is computed by using Eq. (2.25):

$$\Delta t = \frac{-\ln \xi}{W_a + W_d + W_m},\tag{2.32}$$

where ξ is a random number in the $(0, 1)$ interval.

In this simulation, $W_a = 1 \times 10^4/\text{s}$, $W_d = 0$, and $W_m = 0$. Using a random number generator, we get a random number in the $(0, 1)$ interval, $\xi = 0.6154$. According to Eq. (2.32), the real time of the adsorption event is $\Delta t = -\ln \xi/(W_a + W_d + W_m) = 4.855 \times 10^{-5}\text{s}$.

Update the Five Classes

After the execution of a Monte Carlo event, the surface configuration is changed. As a result, the number of nearest neighbors of the surface atoms affected by the

Monte Carlo event is also changed. The five classes and the rates need to be updated accordingly. Note that although the update of the classes can be done by screening the entire lattice, it is more efficient to update the classes locally [155].

In this simulation, the adsorption of the atom on site $(7, 5)$ adds an atom that has zero neighbors so that the number of surface atoms having zero neighbors increases from zero to one. Accordingly, the total desorption rate changes from $W_d = 0$ to $W_d = 4.06/s$ and the total migration rate changes from $W_m = 0$ to $W_m = 233.1/s$. Since the total adsorption rate is site-independent, it remains to be $W_a = 1 \times 10^4/s$. It can be seen that one Monte Carlo event could substantially change the total rates of desorption and migration.

After the real time is advanced by Δt and the five classes are updated, the simulation algorithm goes back to the step of Monte Carlo event selection to start executing the next event. In this way, the simulation continues until it is terminated by the user.

In the remainder of this section, we will also show how to execute desorption and migration events. It can be observed that in the initial stage of the simulation, the total rates of desorption and migration are significantly smaller than the total rate of adsorption. However, the rates of desorption and migration increase significantly when more surface atoms are available due to random adsorption events. To demonstrate how to execute desorption and migration events, we first have multiple atoms adsorbed onto the surface, so that the total rates of desorption and migration are comparable to the total rate of adsorption.

Surface After Multiple Adsorption Events

We continue the Monte Carlo simulation until 30 adsorption events are executed. The resulting surface configuration matrix is shown in Eq. (2.33) and the resulting surface configuration is shown in Fig. 2.4.

$$
S = \begin{pmatrix}
0 & 2 & 0 & 0 & 1 & 1 & 0 & 0 & 1 & 0 \\
0 & 1 & 0 & 0 & 0 & 0 & 0 & 0 & 1 & 0 \\
0 & 0 & 0 & 0 & 0 & 0 & 1 & 0 & 1 & 0 \\
1 & 0 & 0 & 1 & 0 & 0 & 0 & 0 & 0 & 1 \\
0 & 0 & 0 & 2 & 1 & 1 & 0 & 0 & 0 & 0 \\
1 & 0 & 2 & 0 & 0 & 0 & 0 & 1 & 0 & 0 \\
0 & 0 & 0 & 0 & 1 & 1 & 0 & 0 & 0 & 2 \\
0 & 0 & 0 & 0 & 0 & 0 & 0 & 0 & 0 & 1 \\
0 & 1 & 1 & 1 & 1 & 0 & 0 & 0 & 1 & 0 \\
0 & 0 & 0 & 0 & 0 & 0 & 0 & 0 & 0 & 0
\end{pmatrix}.
\tag{2.33}
$$

By grouping all surface atoms based on the number of nearest neighbors, we have that $M_0 = 8$, $M_1 = 14$, $M_2 = 4$, $M_3 = 0$, and $M_4 = 74$. The coordinates of surface atoms in each group are also saved. The total rates are $W_d = 32.48/s$, $W_m = 1864.9$, and $W_a = 1 \times 10^4/s$.

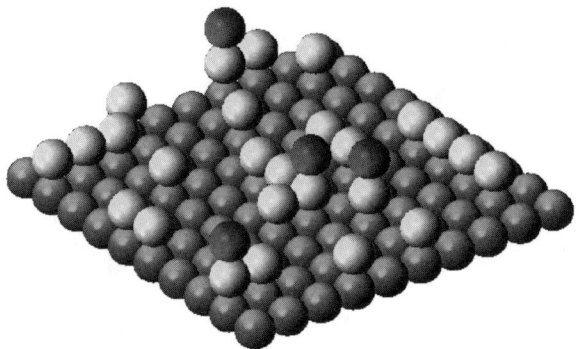

Fig. 2.4. The surface configuration after 30 adsorption events.

Execution of a Monte Carlo Event: Desorption

In the Monte Carlo event selection step, if the generated random number, ζ, satisfies $W_a/(W_a + W_d + W_m) < \zeta < (W_a + W_d)/(W_a + W_d + W_m)$, a desorption event is selected. If the event is a desorption, the kth class in which the desorption event will occur can be selected by finding an integer $k \in \{0, 1, 2, 3, 4\}$ such that

$$\frac{W_a + \sum_{i=0}^{k-1} W_{d_i}}{W_a + W_d + W_m} < \zeta < \frac{W_a + \sum_{i=0}^{k} W_{d_i}}{W_a + W_d + W_m}, \tag{2.34}$$

where W_{d_i} is the total desorption rate of all atoms that have i nearest neighbors. Note that when $k = 0$, Eq. (2.34) becomes

$$\frac{W_a}{W_a + W_d + W_m} < \zeta < \frac{W_a + W_{d_0}}{W_a + W_d + W_m}. \tag{2.35}$$

After the class is selected, a second random number is generated to select the site where the desorption event will be executed.

The current surface configuration is shown in Fig. 2.4, and a random number is generated to select the next Monte Carlo event to be executed. In this simulation, we obtain $\zeta = 0.8413$ for the selection of the next Monte Carlo event. Since ζ satisfies $W_a/(W_a + W_d + W_m) < \zeta < (W_a + W_d)/(W_a + W_d + W_m)$, a desorption event is selected. Then, W_{d_i} is computed by multiplying the desorption rate for a surface atom with i nearest neighbors by the total number of this type of surface atom, M_i. For example, the desorption rate for a surface atom having zero side neighbors is $w_{d_0} = 4.06 \times \exp(-10 \times 0)/\text{s} = 4.06/\text{s}$ and there are $M_0 = 8$ surface atoms that have zero side neighbors. W_{d_0} is, therefore, equal to $4.06 \times 8 = 32.48/\text{s}$. In this way, we also compute that $W_{d_1} = 0.0026/\text{s}$, $W_{d_2} = 3.35 \times 10^{-8}/\text{s}$, $W_{d_3} = 0/\text{s}$, $W_{d_4} = 1.28 \times 10^{-15}/\text{s}$. Based on these rates, the zeroth class, which contains all surface atoms having no nearest neighbors, is selected because ζ satisfies $W_a/(W_a + W_d + W_m) < \zeta < (W_a + W_{d_0})/(W_a + W_d + W_m)$.

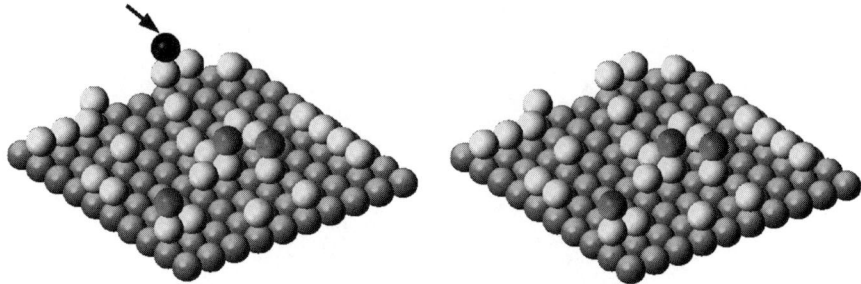

Fig. 2.5. Execution of a desorption event: surface atom selected for desorption (left plot) and the surface configuration after the desorption event (right plot).

In this simulation, the coordinates of the eight surface atoms in the zeroth class are $(1, 2)$, $(3, 7)$, $(5, 4)$, $(6, 1)$, $(6, 3)$, $(6, 8)$, $(7, 10)$, and $(9, 9)$. To randomly select one site among the eight sites in the zeroth class, the eight sites are indexed from 1 to 8 and a random number ζ_1, following the uniform distribution in the $(0, 1)$ interval, is generated to pick a site. The index, i, of the site is determined as $i = ceil(\zeta_1 * 8)$. In this example, we get $\zeta_1 = 0.8099$ and $i = 7$. The site subject to desorption is, therefore, $(7, 10)$. After the site is determined, the desorption event is executed by removing the surface atom on site $(7, 10)$. The surface atom subject to desorption is the black one pointed out by the arrow shown in Fig. 2.5 (left plot); the right plot in the figure shows the surface configuration after the desorption event.

Execution of a Monte Carlo Event: Migration

In the Monte Carlo event selection step, if the generated random number, ζ, satisfies $(W_a + W_d)/(W_a + W_d + W_m) < \zeta < 1$, a migration event is selected. If the event is migration, the kth class in which the desorption event will occur can be selected by finding an integer $k \in \{0, 1, 2, 3, 4\}$ such that

$$\frac{W_a + W_d + \sum\limits_{i=0}^{k-1} W_{m_i}}{W_a + W_d + W_m} < \zeta < \frac{W_a + W_d + \sum\limits_{i=0}^{k} W_{m_i}}{W_a + W_d + W_m}, \qquad (2.36)$$

where W_{m_i} is the total migration rate of all atoms that have i nearest neighbors. Note that when $k = 0$, Eq. (2.36) becomes

$$\frac{W_a + W_d}{W_a + W_d + W_m} < \zeta < \frac{W_a + W_d + W_{d_0}}{W_a + W_d + W_m}. \qquad (2.37)$$

After the class is selected, a second random number is generated to select the site where the migration event will be executed.

Consider again the surface shown in Fig. 2.4; a random number is generated to select the next Monte Carlo event to be executed. In this example, we get a random number $\zeta = 0.9278$ for the selection of the next Monte Carlo event. Since ζ satisfies

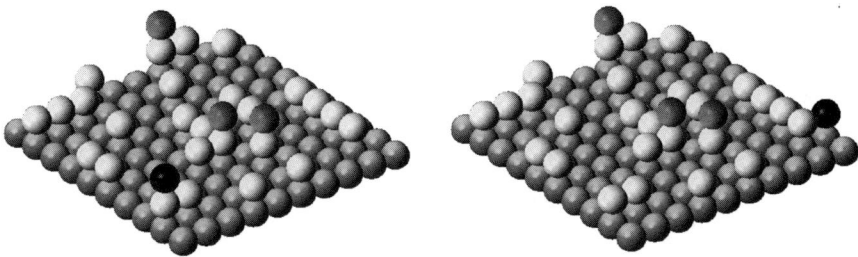

Fig. 2.6. Execution of a migration event: surface atom selected for migration marked as black (left plot) and the surface configuration after the migration event (right plot).

$(W_a + W_d)/(W_a + W_d + W_m) < \zeta < 1$, a migration event is selected. Then, W_{m_i} can be computed by multiplying the migration rate for a surface atom with i nearest neighbors by the total number of this type of surface atom, M_i. For example, the migration rate for a surface atom having zero nearest neighbors is $w_{m_0} = 233.1 \times \exp(-10 \times 0)/s = 233.1/s$ and there are $M_0 = 8$ surface atoms that have zero nearest neighbors. W_{m_0} is, therefore, equal to $233.1 \times 8 = 1864.8/s$. In the same way, we also compute that $W_{d_1} = 0.0026/s$, $W_{d_2} = 3.35 \times 10^{-8}/s$, $W_{d_3} = 0/s$ and $W_{d_4} = 1.28 \times 10^{-15}/s$. Based on these rates, the zeroth class, which contains all surface atoms having no side neighbor, is selected because ζ satisfies $W_a/(W_a + W_d + W_m) < \zeta < (W_a + W_{d_0})/(W_a + W_d + W_m)$.

The coordinates of the eight surface atoms in the zeroth class are $(1, 2)$, $(3, 7)$, $(5, 4)$, $(6, 1)$, $(6, 3)$, $(6, 8)$, $(7, 10)$, and $(9, 9)$. To randomly select one site among the eight sites in the zeroth class, these sites are indexed from 1 to 8 and a random number, ζ_1, following the uniform distribution in the $(0, 1)$ interval, is generated to pick a site. The index, i, of the site is determined as $i = ceil(\zeta_1 * 8)$. In this example, we get $\zeta_1 = 0.0989$ and $i = 1$. The site subject to migration is, therefore, $(1, 2)$. After the site subject to migration is determined, a nearest neighbor that the surface atom hops to also needs to be determined. Since a cubic lattice is used, the surface atom could hop to, four nearest neighbors, which are $(10, 2)$, $(2, 2)$, $(1, 1)$, and $(1, 3)$. Note that since the site $(1, 2)$ is on the boundary of the 10×10 simulation lattice, its nearest neighbor on the left is outside of the simulation lattice. Periodic boundary conditions are applied and its nearest neighbor on the left becomes $(10, 2)$. To randomly select one nearest neighbor among the four nearest neighbors, the four nearest neighbors are indexed from 1 to 4 and a random number, ζ_2, following the uniform distribution in the $(0, 1)$ interval is generated to pick a nearest neighbor to which the surface atom hops. The index, j, of the site is determined as $j = ceil(\zeta_2 * 4)$. In this example, we get $\zeta_2 = 0.1548$ and $j = 1$. The surface atom on site $(1, 2)$ hops to site $(10, 2)$. The migration event is executed by removing the surface atom on the site $(1, 2)$ and adding an atom on site $(10, 2)$. Figure 2.6 (left plot) shows the surface configuration before the migration event; the surface atom selected for migration on site $(1, 2)$ is marked as black. The right plot of the figure shows the surface configuration after the migration event; the surface atom on site $(10, 2)$ (marked as black) is the one that hops from site $(1, 2)$ due to the migration event.

3

Control Using Kinetic Monte Carlo Models

3.1 Introduction

To effectively control material microstructure in multiscale process systems, an appropriate process model for the microscopic phenomena should be incorporated in the control system design. The surface of a thin film is directly formed by surface microprocesses such as adsorption, desorption, and migration. Due to the stochastic nature of the process, the exact time and location of the occurrence of one specific surface microprocess are unknown, and only the probability (rate) with which each surface microprocess may occur is available. The probability that the surface is in a possible configuration is described by the master equation of Eq. (2.11). However, due to the extremely high dimension of a master equation for any system of meaningful size, it is impossible to directly solve the master equation using numerical methods for integration of ordinary differential equations (such as Runge–Kutta). On the other hand, kinetic Monte Carlo (kMC) simulation methods provide a numerical solution to the master equation [80]. The kMC simulation method can be used to predict average properties of the thin film (which are of interest from a control point of view; for example, surface roughness), by explicitly accounting for the microprocesses that directly shape the thin-film microstructure.

In this chapter, we address the real-time feedback control of thin-film growth using kinetic Monte Carlo models. A real-time estimator is first constructed that allows estimates of the surface roughness and growth rate to be computed at a time scale comparable to the real-time evolution of the process. The real-time estimates enable the design of feedback controllers to regulate the thin-film growth process. Methods for the design of single-input–single-output (SISO), multiple-input–multiple-output (MIMO), and predictive controllers using kMC models are then presented. The applications of the developed real-time feedback control methods are illustrated through two thin-film growth processes. The results of this chapter were first presented in [97, 98, 116], and an application of the real-time feedback control methodology to a GaAs deposition process using an experimentally determined kMC process model can be found in [100].

P.D. Christofides et al., *Control and Optimization of Multiscale Process Systems*,
Control Engineering, DOI 10.1007/978-0-8176-4793-3_3,
© Birkhäuser Boston, a part of Springer Science+Business Media, LLC 2009

3.2 Real-Time Estimation

Surface roughness is a property of interest from a control point of view since it directly influences device properties. Various definitions of surface roughness can be found in the literature. Note that estimation and control of surface roughness using different surface roughness definitions can be readily studied within the framework presented in this chapter, which consides the surface roughness represented by the number of broken bonds on the surface and the surface roughness represented by the standard deviation of the surface from its average height [125]. When the surface roughness is represented by the number of broken bonds on the surface, it is computed as follows:

$$r = \frac{\sum(|h_{i+1,j} - h_{i,j}| + |h_{i-1,j} - h_{i,j}| + |h_{i,j+1} - h_{i,j}| + |h_{i,j-1} - h_{i,j}|)}{2 \times N \times N} + 1.$$

$$(3.1)$$

When the surface roughness is represented by the standard deviation of the surface from its average height, it is computed as follows:

$$r = \sqrt{\frac{\sum_{i=1}^{N} \sum_{j=1}^{N} [h_{i,j} - \bar{h}]^2}{N \times N}},$$

$$(3.2)$$

where N is the size of the lattice, $\bar{h}$ is the average height of the surface, and $h_{i,j}$ is the height of the surface at site (i, j).

To achieve control objectives, the real-time acquisition of the controlled variables is important. To this end, an estimation scheme is developed that provides estimates of the surface roughness for all times. The estimation scheme employs kinetic Monte Carlo simulations of the surface together with roughness measurements obtained at discrete-time instants to produce estimates of the surface roughness for all times. The basic idea is to construct a bank of "parallel running" kinetic Monte Carlo simulators of the surface based on small lattice-size models to capture the dominant roughness evolution and utilize the surface roughness measurements to improve upon the predictions of the kinetic Monte Carlo simulators to obtain accurate surface roughness estimates.

3.2.1 Kinetic Monte Carlo Simulator Using Multiple Small Lattices

In the kinetic Monte Carlo simulation, the size of the lattice influences the accuracy of the result and the computational demand. Roughly speaking, the computational complexity of the algorithm we adopt in this work is $O(N^4)$ and the magnitude of the fluctuation in the solution is $O(1/N^2)$, where N is the size of the lattice. The fourth order dependence on computational complexity and the second order dependence of fluctuations on the size of the lattice leave room for reducing the solution time with a relatively small loss of accuracy.

As an illustration, we consider the thin-film growth process presented in Section 2.2 and present simulation results on the evolution of surface roughness under open-loop operation obtained from kMC simulators using different lattice sizes. Both the gas-phase model in Section 2.3 and the surface microstructure model in Section 2.4 are used to simulate the multiscale process, and the two models are coupled through the boundary conditions of the gas-phase PDE model. The process parameters used for the simulations are listed in Table 3.1 and the chamber pressure is 1 atm. The initial precursor concentration above the substrate is zero. In all simulations, the initial surface is a perfect surface. The surface roughness is represented by the number of broken bonds on the surface as defined in Eq. (3.1).

Figure 3.1 shows the evolution of the surface roughness at $T = 600\,\text{K}$ from kinetic Monte Carlo simulators using different lattice sizes: 20×20, 50×50, 100×100, and 150×150. The result from a kinetic Monte Carlo simulation that uses a 20×20 lattice contains significant fluctuations compared to the roughness profiles obtained from a kinetic Monte Carlo simulation that uses a higher-order lattice. However, it gives the same trend of the evolution of surface roughness. Furthermore, the results from kinetic Monte Carlo simulators that use 50×50, 100×100, and 150×150 lattices are very close, thereby implying that a further increase in the lattice size does not improve the accuracy of the results. Due to the algorithm's stochastic nature, it is not possible to obtain the same results from repeated runs starting from the same initial conditions. However, for a sufficiently large lattice size, the results from different runs are consistent in the sense that they provide surface roughness profiles that are very close. Figures 3.2 and 3.3 show the results from three independent kinetic Monte Carlo simulations that utilize a 20×20 lattice and a 50×50 lattice. Clearly, as the lattice size increases, the error among different runs decreases.

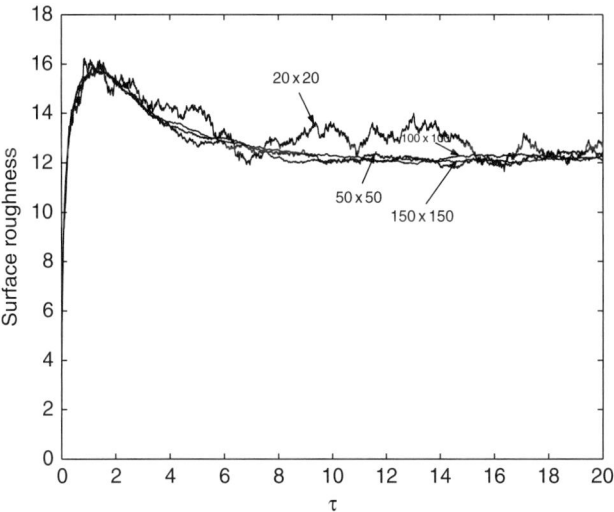

Fig. 3.1. Comparison of the surface roughness profiles obtained from kinetic Monte Carlo simulations that use 20×20, 50×50, 100×100, and 150×150 lattices.

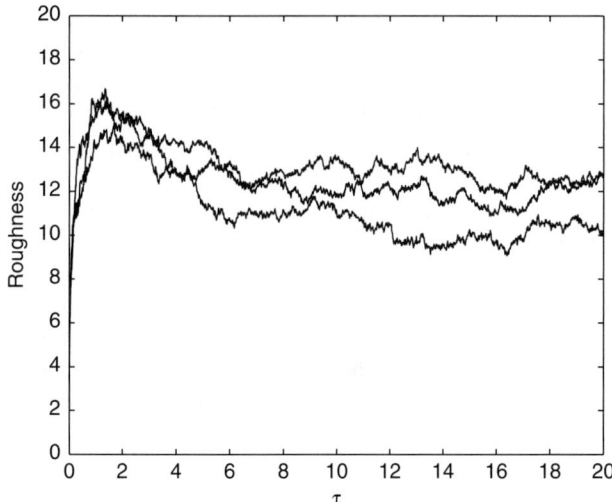

Fig. 3.2. Comparison of the surface roughness profiles from three independent kinetic Monte Carlo simulations that utilize a 20×20 lattice.

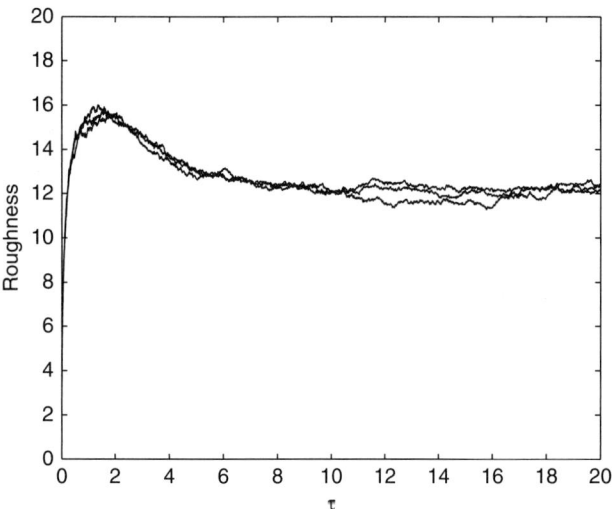

Fig. 3.3. Comparison of the surface roughness profiles from three independent kinetic Monte Carlo simulations that utilize a 50×50 lattice.

Referring to the selection of the lattice size, it is important to point out that the dimension of the small lattice in general should be chosen so that the interactions between the surface atoms are adequately captured and also so that it is large enough to describe all the spatio-temporal phenomena occurring on the surface (e.g., cluster formation). Furthermore, the small lattice should be chosen to provide accurate estimates of the desired properties to be controlled. For example, in the case of surface roughness, this quantity is defined as the average number of broken bonds for every surface atom and the microscopic unit involved is an individual atom. When a 20×20 small lattice is used, the computation of surface roughness involves hundreds of surface atoms, which is adequate to obtain the expected value. However, when the property of interest is, for example, step density, a larger lattice is needed to obtain a convergent average value from the kinetic Monte Carlo simulation. At this point, it is important to note that the proposed reduction of lattice size can be viewed as an alternative (and quite intuitive) way to perform order reduction of the master equation [50].

The fluctuations of the roughness value obtained by using a small lattice Monte Carlo model can be reduced by independently running several Monte Carlo simulations using a small lattice with the same parameters and averaging the roughness values obtained from the different runs. Figure 3.4 shows a comparison of the evolution of surface roughness obtained from (1) a kinetic Monte Carlo simulator that uses a 20×20 lattice, (2) the computation of the average of six independent Monte Carlo simulations that utilize a 20×20 lattice, and (3) a Monte Carlo simulator that uses a 100×100 lattice. These results show that when the outputs from multiple small lattice models are averaged, a more accurate calculation of surface roughness is obtained. However, by increasing the number of Monte Carlo simulations, the computational requirement is also increasing. For the simulations of Fig. 3.4, the computational time for the averaged roughness is six times higher than that needed to perform one Monte Carlo simulation run and approximately equal to the computational time needed to

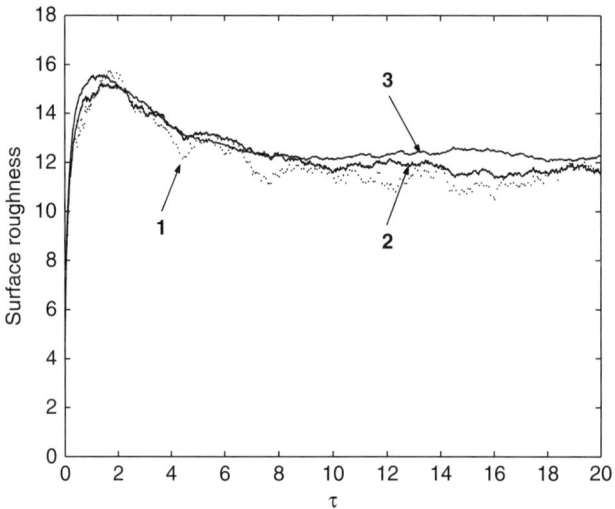

Fig. 3.4. Comparison of the evolution of surface roughness from (1) a Monte Carlo simulation that uses a 20×20 lattice, (2) the computation of the average of six independent Monte Carlo simulations that utilize a 20×20 lattice, and (3) a Monte Carlo simulation that uses a 100×100 lattice.

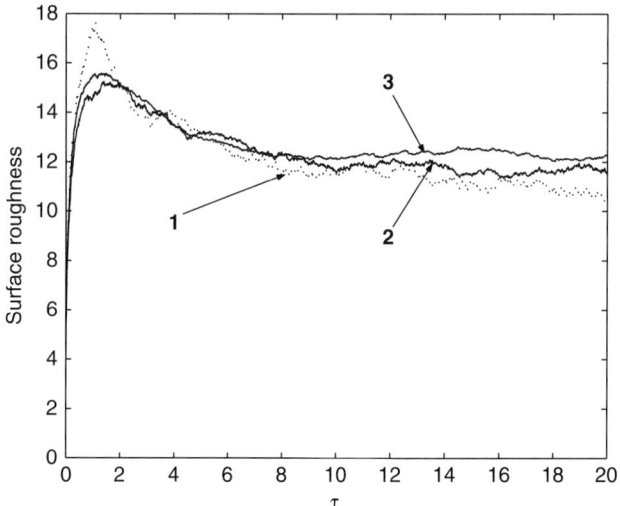

Fig. 3.5. Comparison of the evolution of surface roughness from (1) a Monte Carlo simulation that uses a 30×30 lattice, (2) the computation of the average of six independent Monte Carlo simulations that utilize a 20×20 lattice, and (3) a Monte Carlo simulation that uses a 100×100 lattice.

run a Monte Carlo simulation on a 30×30 lattice. Figure 3.5 shows a comparison of the evolution of roughness obtained from: (1) a Monte Carlo simulator that uses a 30×30 lattice, (2) the computation of the average of six independent Monte Carlo simulations that utilize a 20×20 lattice, and (3) a Monte Carlo simulator that uses a 100×100 lattice. The results of Fig. 3.5 show that the roughness obtained by averaging six Monte Carlo simulations, which use a 20×20 lattice model, is closer to that obtained from the 100×100 lattice model and is superior to that obtained from the 30×30 lattice model.

3.2.2 Adaptive Filtering and Measurement Compensation

The predicted profiles of surface roughness, which are obtained from kinetic Monte Carlo simulation based on multiple small lattice models, still contain stochastic fluctuations and are not robust (due to the open-loop nature of the calculation) with respect to disturbances and variations in process parameters. To alleviate these problems, we combine the small lattice kinetic Monte Carlo simulators with an adaptive filter to reject the stochastic fluctuations, and a measurement error compensator to improve the surface roughness estimate using online measurements.

To overcome the fluctuations introduced by the smaller lattice size, a second-order linear filter is used to reject the noise in the roughness values obtained from the kinetic Monte Carlo simulator that uses multiple small lattice models with the following state-space representation:

$$\frac{d\hat{y}_r}{d\tau} = y_1,$$

$$\frac{dy_1}{d\tau} = \frac{K}{\tau_I}(y_r - \hat{y}_r) - \frac{1}{\tau_I}y_1, \qquad (3.3)$$

where y_1 is a state of the filter, y_r is the output of the kinetic Monte Carlo simulator that uses multiple small lattice models, $\hat{y}_r$ is the filter output, K is the filter gain and τ_I is the time constant. To accelerate the response of the filter and avoid overshoot, $\tau_I = 0.5/K$.

Due to the Arrhenius-like dependence of the rate of desorption and surface migration on temperature, the dynamics of the roughness with respect to temperature variations is very fast. To this end, we need a filter that can both track the fast dynamics and reject the noise so that the noise will not deteriorate the controller's performance. However, fast tracking and efficient noise rejection are two conflicting objectives that are very hard to achieve simultaneously. Fortunately, during the fast dynamic stage, the roughness is very large compared to the fluctuations and the effect of the fluctuations on the controller performance is insignificant. The fluctuations begin to deteriorate the controller's performance significantly only when the controlled roughness is close to the set-point value. Motivated by this, an adaptive scheme is used to determine the gain of the filter online such that the filter focuses on tracking the growth dynamics at the fast dynamic stage and on noise rejection when the surface dynamics slow down. To achieve this, the gain of the filter is adjusted according to the following law:

$$K(\tau) = K_0 \frac{|\int_{\tau-\Delta\tau}^{\tau} y_r(t)dt - \int_{\tau-2\Delta\tau}^{\tau-\Delta\tau} y_r(t)dt|}{\Delta\tau^2} + K_s, \qquad (3.4)$$

where K_0 is a constant, K_s is the steady-state gain for the adaptive filter, and $\Delta\tau$ is the dimensionless time interval between two updates of K. To make the filter focus on tracking the growth dynamics at the fast dynamic stage and on noise rejection when the surface dynamics slow down, the K_0 and K_s in Eq. (3.4) are tuned such that the first term on the right-hand side of Eq. (3.4) is dominant during the fast dynamics stage and the second term on the right-hand side of Eq. (3.4) is dominant when the dynamics slow down. The filter parameters (K_0 and K_s) can be determined from a step test of the process. Regarding the choice of $\Delta\tau$, although a better tracking performance is expected when a smaller $\Delta\tau$ is used, a very small $\Delta\tau$ will introduce the effect of fluctuations on the filter gain and should be avoided.

The significant fluctuation is not the only problem that must be fixed when a small lattice is used in the kinetic Monte Carlo simulator. There is also model inaccuracy when the outputs of kinetic Monte Carlo simulators that use multiple small lattices and a high-order lattice are compared. This can be corrected by using a measurement error compensator that uses the available roughness measurements obtained from a kinetic Monte Carlo simulator that uses the high-order lattice model (or directly from experiments at discrete time instants) to produce an improved roughness estimate. The state-space representation of the measurement error compensator is

$$\frac{de}{d\tau} = K_e(y_h(\tau_{m_i}) - \hat{y}(\tau_{m_i})), \quad \tau_{m_i} < \tau \leq \tau_{m_{i+1}}, \quad i = 1, 2, \ldots, \qquad (3.5)$$

and the final roughness estimates are computed by

$$\hat{y} = \hat{y}_r + e. \tag{3.6}$$

In the above equations, K_e is the compensator gain, e is the estimated model error, which is used to compensate the model output, $\hat{y}$ is the roughness estimate, $\hat{y}_r$ is the filtered output from a kinetic Monte Carlo simulator that uses multiple small lattices, and y_h is the output of a kinetic Monte Carlo simulator that uses the high-order lattice (in an experimental setup, y_h could be obtained from the measurement sensor). Since the roughness measurements are only available at discrete points in time, $\tau_m = [\tau_{m_1}, \tau_{m_2}, \ldots]$, the right-hand side of Eq. (3.5) is computed at the time a roughness measurement is available and is kept in this value in the time interval between two available roughness measurements.

The combination of the adaptive filter and the measurement error compensator functions as a roughness estimator, which can accurately predict the evolution of roughness during the thin-film growth by using measurements of the precursor concentration above the substrate and surface roughness. In this development, we assume that the estimate of precursor concentration above the substrate is available; the surface precursor concentration can be estimated using estimation methods for continuum-type PDE models [28].

3.3 Feedback Control Design

The developed real-time estimator provides a computationally feasible approach to predict the evolution of important process variables. In this section, design methods for SISO control, MIMO control, and model predictive control using the real-time estimator are presented.

3.3.1 An Estimator/Controller Structure

SISO Control

We initially consider the use of a proportional-integral (PI) feedback controller to regulate the surface roughness by manipulating the substrate temperature. Specifically, the controller has the following form:

$$u(\tau) = K_c[(\hat{y} - y_{\text{set}}) + \frac{1}{\tau_c} \int_0^\tau (\hat{y} - y_{\text{set}})dt], \tag{3.7}$$

where y_{set} is the set-point of the output, $\hat{y}$ is the output of the roughness estimator, K_c is the proportional gain, and τ_c is the integral time constant.

The PI controller is coupled with the roughness estimator presented in the previous section. A diagram of the closed-loop system under the proposed estimator/controller structure is shown in Fig. 3.6. In this structure, the estimator, which includes multiple kinetic Monte Carlo simulators using small lattice models, an adaptive filter, and a measurement error compensator, is used to provide estimates of the

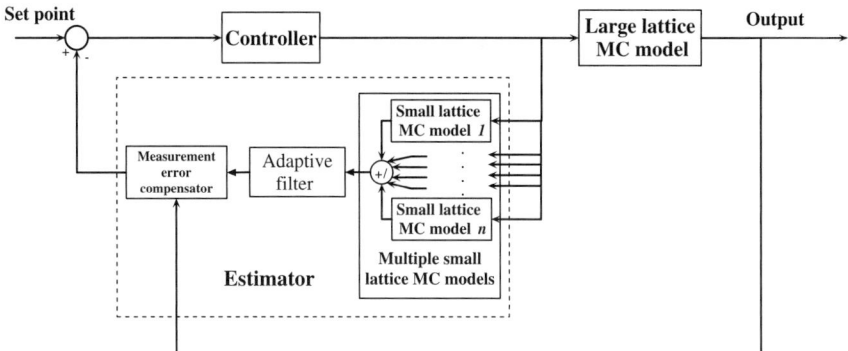

Fig. 3.6. Diagram of the estimator/controller structure using a kinetic Monte Carlo simulator based on multiple small lattice models.

controlled variable (surface roughness or growth rate) in a time scale comparable to the real-time evolution of the process. The estimates are used in the controller to determine the control action.

MIMO Control

We now turn our attention to the design and evaluation of a multivariable feedback control structure, based on kinetic Monte Carlo models, used to control the surface roughness and thin-film growth rate by manipulating the substrate temperature and inlet precursor concentration. A multivariable feedback control structure is developed that explicitly compensates for the effect of the multivariable interactions occurring in the process. The controller structure is obtained by introducing a compensation block between the multiple single-loop controllers and the process. A diagram of the multivariable control system using the estimator/controller structure with interaction compensation is shown in Fig. 3.7. $G_1(s)$ is the transfer function between the substrate temperature and the growth rate and $G_2(s)$ is the transfer function between the inlet precursor mole fraction and the growth rate. Step tests are used to identify the expression and parameters of $G_1(s)$ and $G_2(s)$.

3.3.2 Model-Predictive Control

In order to achieve a robust closed-loop operation in certain highly complex deposition processes, a kinetic Monte Carlo model-based predictive control scheme is developed. Figure 3.8 shows the block diagram of the closed-loop system. A reference trajectory of the instantaneous surface roughness of the thin film is selected based on off-line optimization, and in this work for simplicity, the profile of the surface roughness of the thin film in an ideal open-loop deposition (no disturbance is assumed to affect the process and the final surface roughness is taken to be the desired value) is chosen. Using such a reference trajectory, instead of solving the receding horizon optimization problem of minimizing the difference between the final surface roughness and the desired value with multiple decision variables, we only need to

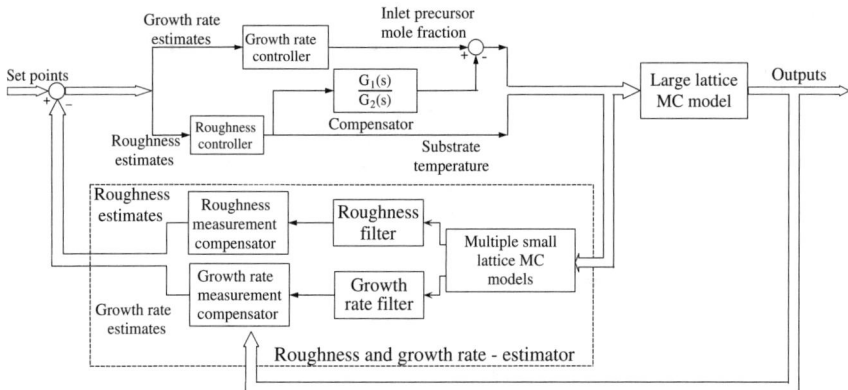

Fig. 3.7. Diagram of multivariable feedback control system with interaction compensation.

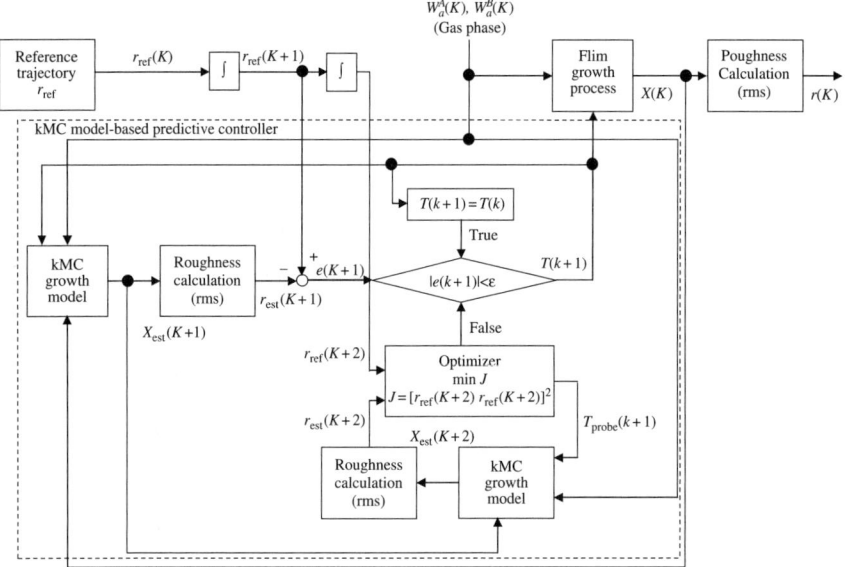

Fig. 3.8. Block diagram of the closed-loop system with the kMC model-predictive controller.

solve the fixed short-horizon optimization problem of minimizing the difference between the instantaneous surface roughness and the reference value with a single decision variable. Therefore, the computation time of each optimization is greatly reduced, since the kMC simulation duration is reduced from the scale of the total deposition time to the controller turnover time. This is very important since kMC simulation is relatively time-consuming and large-scale numerical optimization using kMC models is almost impossible to solve in real time.

During each control cycle, the surface configuration $X(k)$ is first measured. An estimate of the surface configuration at the next control action time $X_{\text{est}}(k+1)$ is computed based on the current process conditions using the proposed kMC model, and the estimated surface roughness value $r_{est}(k+1)$ is compared with the reference value $r_{ref}(k+1)$. If the error is less than ε, the next controller output $T(k+1)$ is set to be the same as the current output $T(k)$. If the error is larger than ε, the optimizer is called to compute the output value of the next control action $T(k+1)$ so that the error between the surface roughness after the next control action $r(k+2)$ and the reference value $r_{ref}(k+2)$ is minimized.

The optimizer uses direct search to find the optimal solution since the kMC model does not have a closed-form expression. The estimate of the surface roughness after the next control action $r_{est}(k+1)$ is computed using the proposed kMC model based on the estimated surface configuration before the next control action $X_{est}(k+1)$, the probe output value $T_{probe}(k+1)$, and current process conditions. When the search precision is 1 K, the optimization problem can be solved by an entry-level personal computer within the controller turnover time of 10 s. Furthermore, the speed and the precision of the direct search optimization algorithm can be substantially improved by parallel computing.

Remark 3.1. It is important to comment on the dependence of the steady-state surface roughness on the lattice size of the kMC models of various deposition processes. In the most simplified deposition process model where only random deposition events take place, the expected surface roughness keeps increasing as time increases, since there are no correlations between any two lattice sites and the fluctuations in the heights obey a Poisson process. However, when surface relaxation and/or diffusion take place in the process, the surface profiles tend to be smoothened out and the expected surface roughness reaches a steady-state value at large times. The steady-state values of the roughness depend on the lattice size differently in different dimensions of the models, which has been verified by both numerical simulations and theoretical derivations. Specifically, for a 2D model as in this work, i.e., the lattice sizes are predefined in the x and y directions, the steady-state surface roughness is weakly dependent on the lattice size [72, 81] for sufficiently large lattice sizes. Thus, it allows controller design based on smaller lattice size models. For a 1D model of the surface roughness of a random deposition with surface relation (RDSR) process, the dependence of roughness on the lattice size is stronger. There is a roughness exponent of 0.5 in this case, i.e., the steady-state surface roughness is proportional to the order of square root of the lattice size [40]. These results are also consistent with the solutions of the Edwards and Wilkinson equations [39] in 1D and 2D, which are stochastic PDEs that can be derived from the master equations of the kMC models of RDSR processes.

3.4 Application to a Thin-Film Growth Process

In this chapter, we present simulation results on applications of the SISO and MIMO feedback controllers to the thin-film growth process presented in Section 2.2 with process parameters presented in Table 3.1. The process is the growth of a thin film

Table 3.1. Parameters of the thin-film growth process.

Sticking coefficient	s_0	0.1	
Precursor molecular weight	m	4.65×10^{-26}	kg
Surf. site concentration	C_{tot}	10^{19}	sites/m^2
Event-freq. const	k_{d0}	10^{13}	s^{-1}
Bond energy	E	0.74	eV
Desorption energy	E_d	0.74	eV
Migration energy	E_m	0.44	eV
Strain rate	a	5	s^{-1}

from a fluid in a vertical, stagnation-flow geometry as shown in Fig. 2.1. In this geometry, the inlet fluid flow forms a uniform boundary layer adjacent to the surface of the substrate and precursor atoms diffuse through such a boundary and deposit a uniform thin film [49]. Upon arrival at the surface, the precursor atoms are adsorbed onto the surface. Subsequently, adsorbed atoms may desorb to the gas-phase or migrate on the surface. The gas-phase model in Section 2.3 and the surface microstructure model in Section 2.4 are used for simulations. The open-loop dynamics of the thin-film growth process are shown in Section 3.2.1.

3.4.1 SISO Control: Surface Roughness Regulation

In this case study, we consider as manipulated variable the substrate temperature, which is assumed to change only with respect to time. This is a reasonable formulation for the manipulated input and is practically feasible for many experimental and industrial deposition processes. With such a manipulated input formulation, the only variable that can be controlled is a spatially averaged roughness. The surface roughness defined in Eq. (3.1) is used. We could have formulated a control problem under the assumption that a large number of manipulated inputs (control actuators) are available to control surface roughness with higher precision, but such a control problem formulation is not considered at the present time but will be discussed in Chapters 4 and 5. As we will see later, it is possible by manipulating the substrate temperature (single-input formulation) to achieve an overall very smooth surface configuration. In this case study, the process parameters are shown in Table 3.1 and the chamber pressure is 1 atm.

Real-Time Roughness Estimator

A roughness estimator, which combines a kMC simulator using multiple small lattice models, a second-order adaptive filter, and a measurement compensator, is used to accurately predict the evolution of roughness during the thin-film growth in real time. The size of the small lattice has to be selected to make the model computations be at a time scale comparable to the process real-time evolution. Based on the simulation results shown in Figs. 3.1 to 3.5, when the size of the lattice is reduced to 20×20, the time of simulation is comparable to the real-time process evolution. A kMC simulator

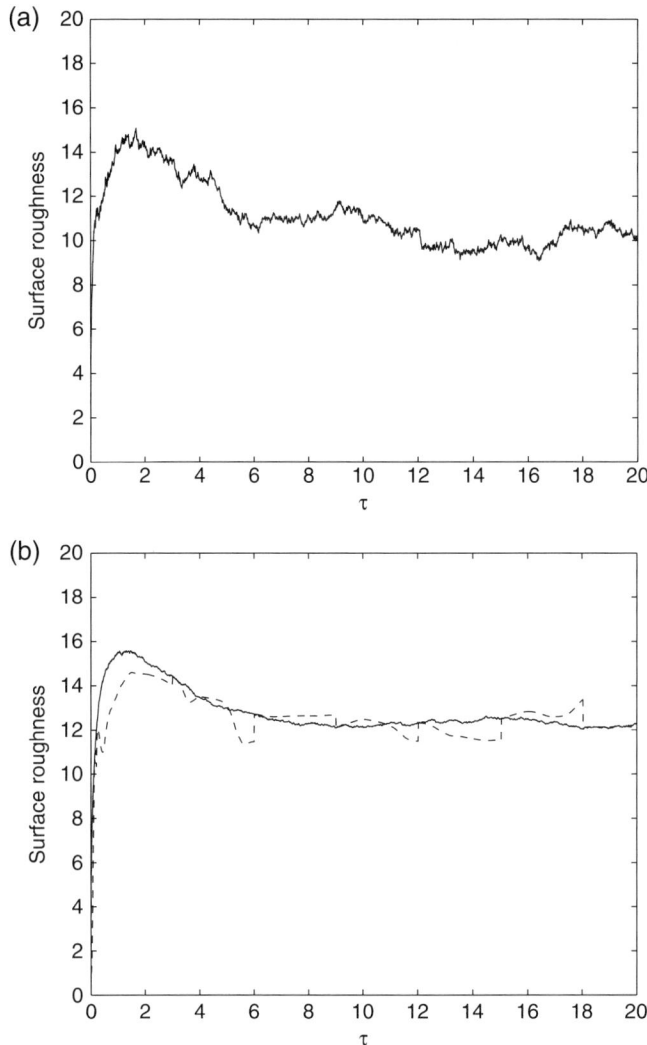

Fig. 3.9. (a) plot: surface roughness profile obtained directly from a kMC simulation that uses a 20 × 20 lattice; (b) plot: surface roughness profile from the roughness estimator (dashed line) and from a kMC simulation that uses a 100 × 100 lattice (solid line).

using a 100 × 100 lattice is used to simulate the process. The real-time roughness estimator is designed with the parameters shown in Table 3.2 (column SISO).

The left plot of Fig. 3.9 shows the roughness obtained directly from a kMC simulator using a 20 × 20 lattice, which contains significant stochastic noise. The right plot illustrates the prediction of surface roughness by the roughness estimator and its comparison to the output from the kMC model, which uses a large lattice (100 × 100). The results clearly show that when measurements are used, the estimator based on a kMC simulator that uses a small lattice can predict the evolution of surface roughness, while the computational requirements are kept within the limit that online control is possible.

Feedback Control of Surface Roughness

A closed-loop simulation was run to evaluate the effectiveness of the estimator/controller structure shown in Fig. 3.6. The size of the small lattice is 20×20, and a 50×50 lattice Monte Carlo model is used to describe the evolution of the process. The desired roughness is 1.5. The time interval between two available measurements is 0.3 s; this is consistent with spectroscopic ellipsometry techniques that can be used to measure surface roughness online [159]. The controller parameters used in the simulation are shown in Table 3.2 (column SISO).

Initially, the thin film grows on a perfect surface at $T = 600$ K and the roughness increases. Then, the controller is activated when the roughness reaches 15.5. Figure 3.10 shows the evolution of surface roughness and the profile of the substrate temperature under feedback control. The results clearly show that the proposed estimator/controller structure can successfully drive the surface roughness to the desired set-point value.

The microconfiguration of the surface before the controller is activated and at the end of the closed-loop simulation run is shown in Fig. 3.11. The controller successfully reduces the surface roughness of the thin film.

Advantages of the Estimator/Controller Structure

To show the importance of using the roughness estimator for feedback control and not relying exclusively on the roughness measurements (which are obtained every 0.3 s), we applied the PI controller (with the same parameters as in column SISO of Table 3.2) to the multiscale process model assuming that new roughness measurements are fed into the controller every 0.3 s (this is consistent with our previous simulations). Also, to prevent the substrate temperature from increasing to an unreasonably high value, the substrate temperature is constrained to be below $T_{\max} = 1100$ K. We note that when the roughness is controlled using the proposed controller/estimator structure, the substrate temperature is always lower than $T_{\max}$ (see Fig. 3.10). Figure 3.12 shows the evolution of surface roughness and substrate temperature. Due to the discrete measurements, we observe significant oscillations in the surface roughness profile; this simulation shows the usefulness of the proposed estimator/controller structure.

We note that the fundamental reason for the poor performance shown in Fig. 3.12 is that the discrete measurements cannot capture the full dynamics of the system, which prevents the controller from computing efficient control actions. Tuning the controller cannot achieve a control performance as good as that achieved under feedback control using the roughness estimator. In [97], various tuning schemes are applied to the PI controller without using the real-time roughness estimator, and simulation results are presented. When the controller is tuned to achieve fast dynamics,

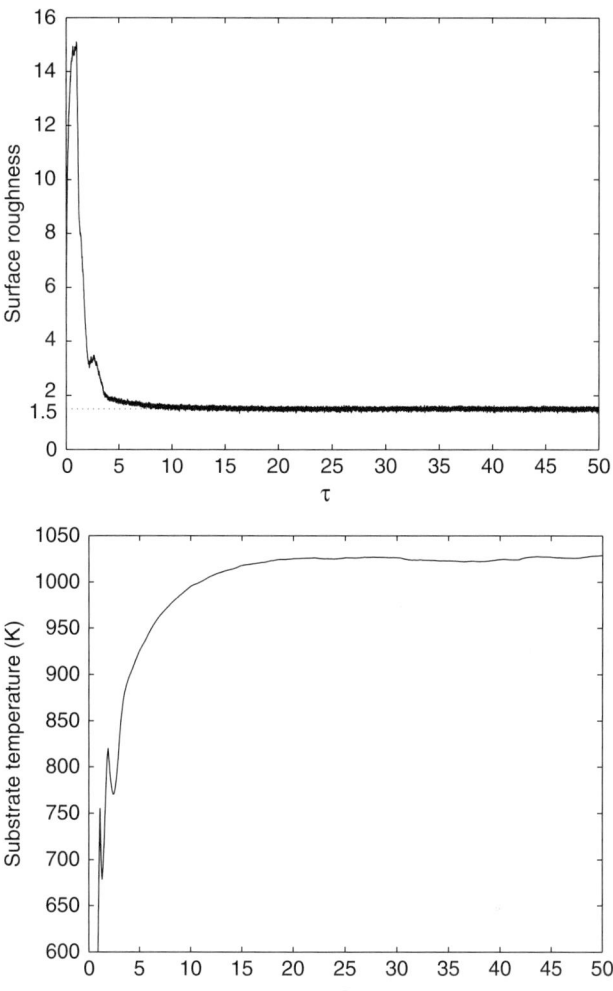

Fig. 3.10. Evolution of the surface roughness and the substrate temperature under feedback control.

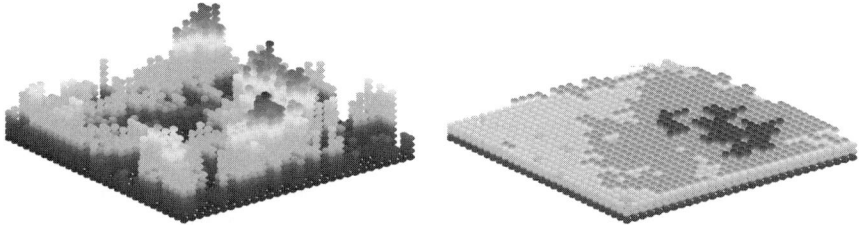

Fig. 3.11. The microconfiguration of the surface for $T = 600$ K (left plot) and that at the end of the closed-loop simulation run (right plot).

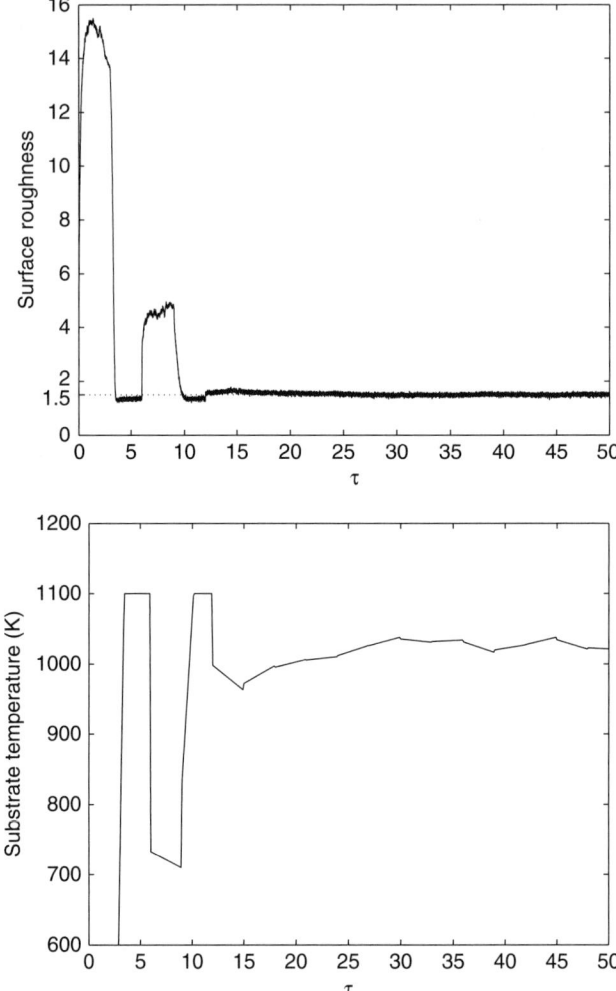

Fig. 3.12. Evolution of the surface roughness and substrate temperature under feedback controller without roughness estimator.

it will result in significant oscillations in the roughness output. On the other hand, when the controller is tuned to avoid oscillations, a much longer time is required for the process to reach the steady state. We tried different parameters of the PI controller, but it turns out it is hard to achieve short transient time and less oscillation simultaneously when the surface roughness is controlled exclusively by relying on the discrete measurements. To compare the performance of the various control approaches, the roughness profiles obtained from different control schemes are shown in Fig. 3.13. It is clear that the proposed estimator/controller scheme results in the best closed-loop performance.

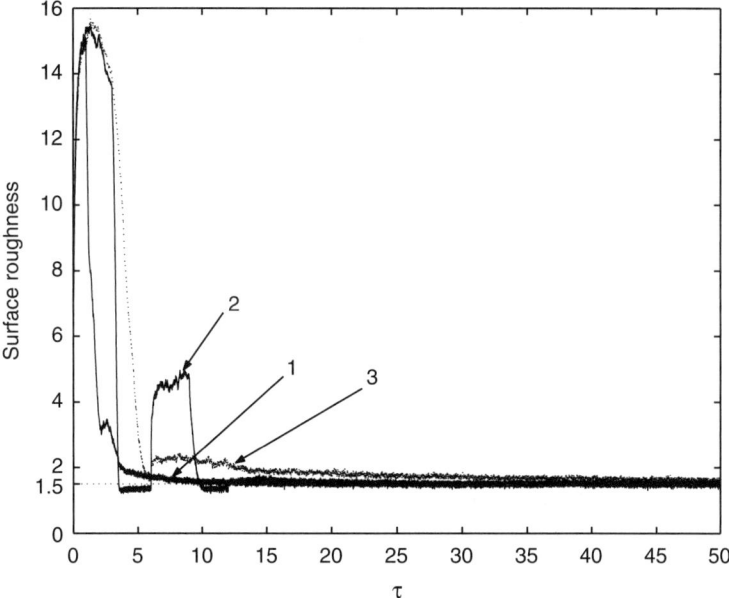

Fig. 3.13. Comparison of surface roughness profiles: (1) roughness profile under feedback control using roughness estimator; (2) roughness profile under feedback control without using roughness estimator with controller parameters shown in column SISO of Table 3.2; and (3) roughness profile under feedback control without using roughness estimator with controller parameters tuned to avoid significant oscillation.

Other advantages of the proposed estimator/controller structure for surface roughness regulation are also demonstrated in [97]. When there is a time delay in the measurements of the surface roughness, the PI controller exclusively relying on discrete measurements (with parameters shown in column SISO of Table 3.2) is not able to drive the surface roughness to the set-point value. On the other hand, in the estimator/controller structure, the measurement delay has very little effect on the closed-loop performance. Furthermore, we have shown that the estimator/controller structure is able to control the surface roughness independently of the frequency at which the roughness measurements are available. The robustness properties of the estimator/controller structure were also tested. In [97], we considered controlling the thin-film growth process in the presence of 10% uncertainty in the energy associated with a single bond on the surface. Simulation results showed that the controller exhibited very good robustness properties in the presence of this parametric uncertainty.

3.4.2 MIMO Control: Surface Roughness and Growth Rate Regulation

The efficient production of high-quality thin films requires that the surface roughness and growth rate be maintained at desired levels. Therefore, the objectives of this section are to study the nature of the multivariable control problem and to demonstrate the effectiveness of the multivariable feedback control structure shown in Fig. 3.7. The feedback control system makes use of the real-time estimator, which provides a computationally feasible approach to predict the growth rate and surface roughness of the thin film in real time. Given the set of available manipulated inputs and the desired control objectives, the control problem is formulated as that of regulating the surface roughness and growth rate by manipulating the substrate temperature and precursor mole fraction in inlet gases. The surface roughness defined in Eq. (3.1) is used. We will begin with a study of the feasibility of the control problem formulation, continue with an analysis and evaluation of the effect of input–output interactions on closed-loop performance, and close with the design and evaluation of the multivariable feedback control structure shown in Fig. 3.7. In this case study, the process parameters are shown in Table 3.1 and the chamber pressure is 10 Pa.

KMC Simulator Based on Multiple Small Lattice Models

The growth rate, gr (units monolayer/second, ML/s), determines how fast the thin film grows. In this study, the growth rate is defined as the difference between the rate of adsorption and the rate of desorption on the surface and is given by the following expression:

$$gr = \frac{\sum_{i,j} r_{ai,j} - r_{di,j}}{N \times N}, \qquad (3.8)$$

where $r_{ai,j}$ and $r_{di,j}$ (units s^{-1}) are the rates of adsorption and desorption at the site (i, j) of the surface, respectively, and N is the size of the lattice. The rate $r_{di,j}$ depends on the number of nearest neighbors at site (i, j) and is obtained from the kinetic Monte Carlo simulations.

To implement real-time feedback control based on a model that captures the evolutions of surface roughness and growth rate, the size of the lattice has to be selected to make the model solution time comparable to the process real-time evolution, while capturing the dominant phenomena occurring on the surface. In our simulations, when the size of the lattice is reduced to 30×30, the solution time of the kinetic Monte Carlo simulation is comparable to the real-time process evolution and the average values of the surface roughness and growth rate approximate well the average values of these variables, which are obtained by running the kinetic Monte Carlo simulation on a 120×120 lattice (this is a sufficiently large lattice to ensure simulation results that are independent of the lattice size; see the simulation results shown in Fig. 3.17).

However, the outputs from a kMC simulation using a 30×30 lattice contain significant stochastic fluctuations, thus, they cannot be directly used for feedback control [such an approach would result in significant fluctuations of the control action, which could perturb unmodeled (fast) process dynamics and should be avoided]. The fluctuations on the values of the outputs (i.e., surface roughness and growth rate) obtained from the kMC simulation using the 30×30 lattice can be reduced by independently running several small lattice kMC simulations with the same parameters and averaging the outputs of the different runs. Approximately, the computational

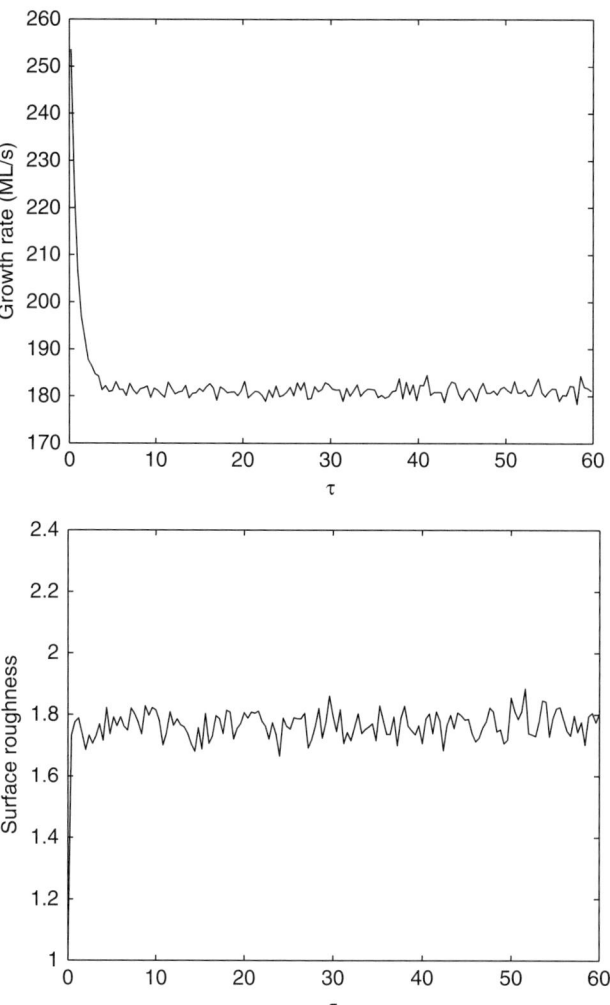

Fig. 3.14. Growth rate and surface roughness from the computation of the average of six independent kinetic Monte Carlo simulations that utilize a 20 × 20 lattice.

requirement for a kMC simulation using a 30 × 30 lattice is close to that for six independent kMC simulations that utilize a 20 × 20 lattice. Figure 3.14 shows the growth rate and surface roughness obtained from the computation of the average of six independent kMC simulations that utilize a 20 × 20 lattice. Figure 3.15 shows the growth rate and roughness profiles obtained from a kMC simulation that uses a 30 × 30 lattice. Comparing the simulation results shown in Figs. 3.14 and 3.15, we find that the roughness and growth rate obtained by averaging six independent kinetic Monte Carlo simulations, which use a 20 × 20 lattice model, contain fewer fluctuations than those obtained from a kinetic Monte Carlo simulation that uses a

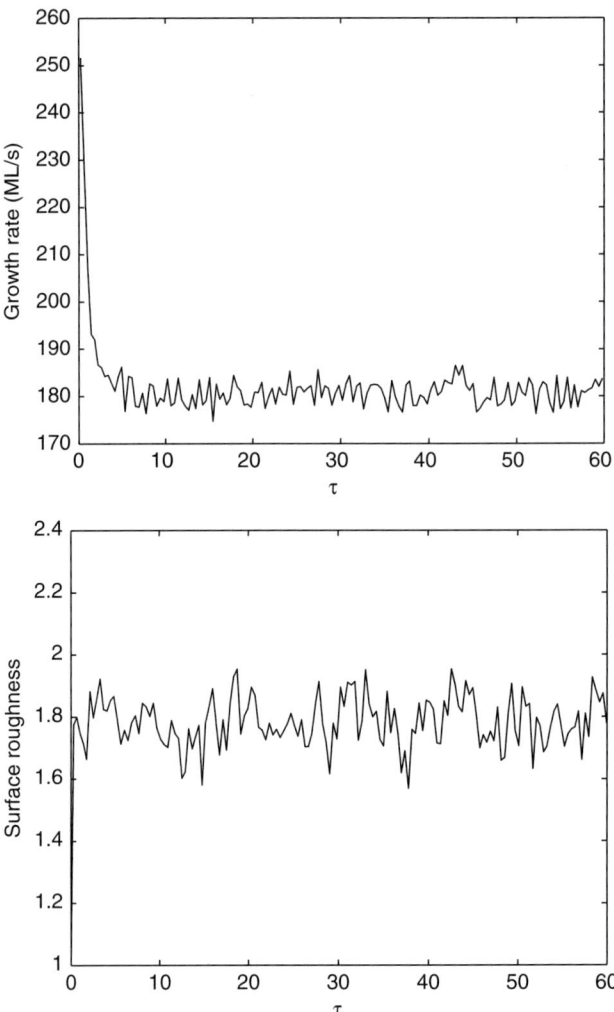

Fig. 3.15. Growth rate and surface roughness from a kinetic Monte Carlo simulation that uses a 30×30 lattice.

30×30 lattice model. This demonstrates that, for a fixed computational time, the use of a kMC simulator based on multiple small lattice models would yield growth rate and surface roughness profiles with fewer fluctuations compared to a kMC simulator that uses a single lattice with a larger size.

Real-Time Surface Roughness and Growth Rate Estimator

The predicted profiles of surface roughness and growth rate, which are obtained from kMC simulation based on multiple small lattice models, still contain stochastic fluctuations and are not robust (due to the open-loop nature of the calculation) with respect to disturbances and variations in process parameters. To alleviate these problems, we combine the small lattice kMC simulator with an adaptive filter, to reject the stochastic fluctuations on the surface roughness and growth rate profiles, and a measurement error compensator to improve the estimates of these variables using online measurements. We use the same adaptive filter and measurement error compensator structure for both the surface roughness and the growth rate. The measurement error compensator uses the available online measurements (in the numerical simulations the values of the surface roughness and growth rate obtained from the large lattice model are used) to produce improved estimates of the surface roughness and growth rate.

Figure 3.16 shows the growth rate and surface roughness profiles computed by the estimator, which uses a kinetic Monte Carlo simulator based on six 20×20 lattice models (solid lines); they are compared with the growth rate and surface roughness profiles obtained from a kinetic Monte Carlo simulator that uses a 120×120 lattice model. The results clearly show that the developed estimator can accurately predict the evolution of the growth rate and surface roughness. Note also that the developed estimator can be used for real-time feedback control since the computational time needed to run a kinetic Monte Carlo simulation based on six 20×20 lattice models is comparable to the real-time process evolution.

Referring to the selection of the lattice size, when the surface roughness and growth rate are considered in the process with parameters shown in Table 3.1, we found that an 80×80 lattice is sufficient to capture the evolution of the process and that a further increase in the lattice size leads to no observable improvement in the accuracy of the simulation results; this is shown in Fig. 3.17, which shows comparisons of surface roughness profiles and growth rate profiles from a kinetic Monte Carlo simulator that uses an 80×80 lattice and those from a kinetic Monte Carlo simulator that uses a 120×120 lattice. Therefore, in the remainder of this work, we use a kinetic Monte Carlo simulator that uses an 80×80 lattice to describe the evolution of the thin-film growth under open-loop and closed-loop conditions.

Feasibility of Simultaneous Control of Surface Roughness and Growth Rate

In this subsection, we establish that it is feasible to control the surface roughness and growth rate by manipulating the substrate temperature and precursor mole fraction in inlet gases. To this end, we perform the following set of closed-loop simulations: (1) The growth rate is controlled by manipulating the precursor mole fraction in the inlet gas while the substrate temperature is kept constant; and (2) the surface roughness is controlled by manipulating the substrate temperature while the inlet precursor mole fraction is kept constant. In our calculations, since the precursor mole fraction is very small, we assume that the chamber pressure is independent of the precursor mole fraction. In each simulation, a controller/estimator structure of the type shown in Fig. 3.6 is used to control the process. In this structure, the estimator, which includes multiple kinetic Monte Carlo simulators using small lattice models, an adaptive filter, and a measurement error compensator, is used to provide estimates of the

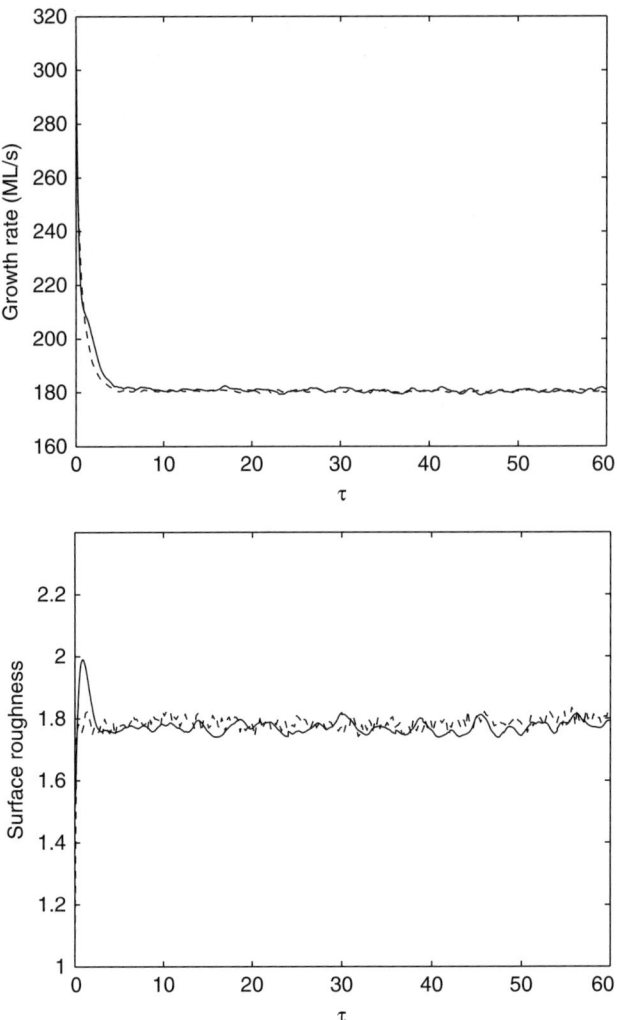

Fig. 3.16. Growth rate and surface roughness profiles from the estimator (solid lines) and from a kinetic Monte Carlo simulation that uses a 120×120 lattice model (dashed lines).

controlled variable (surface roughness or growth rate) in a time scale comparable to the real-time evolution of the process. The estimates are used in the controller to determine the control action. Since the models that describe the evolution of the surface roughness and growth rate are not available in closed form, a PI controller is used to compute the control action:

$$u(\tau) = K_c \left[(y_{\text{set}} - \hat{y}) + \frac{1}{\tau_c} \int_0^\tau (y_{\text{set}} - \hat{y}) dt \right], \tag{3.9}$$

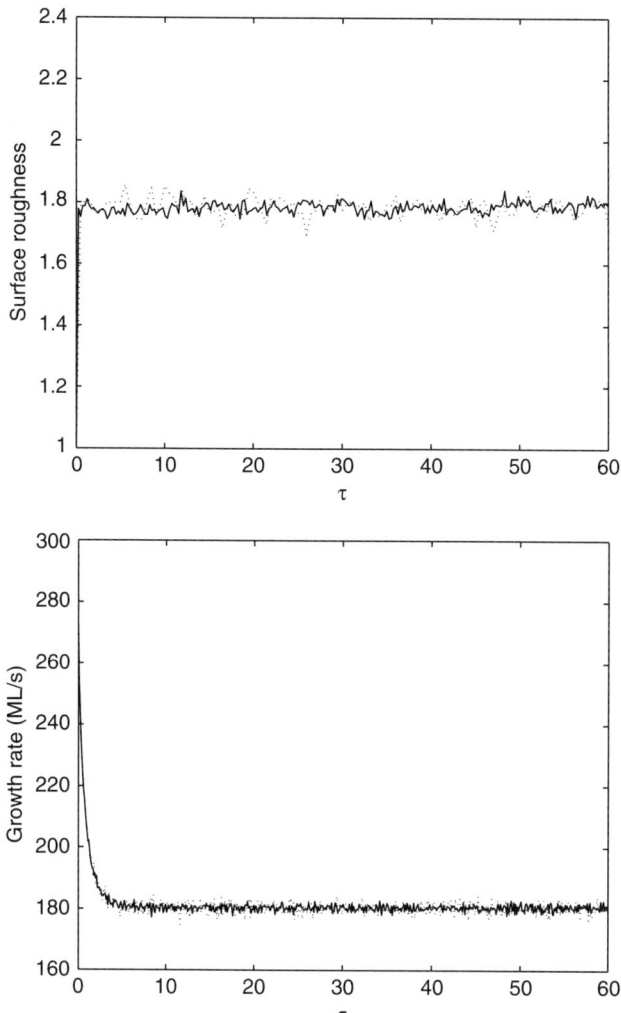

Fig. 3.17. Comparisons of the surface roughness and growth rate profiles from a kinetic Monte Carlo simulation that uses an 80×80 lattice (dotted) and those from a kinetic Monte Carlo simulation that uses a 120×120 lattice (solid).

where y_{set} is the set point of the output, $\hat{y}$ is the output of the estimator, K_c is the proportional gain, and τ_c is the integral time constant.

When the growth rate is controlled, the size of the small lattice is 20×20 and the outputs of six small lattice kinetic Monte Carlo simulators are averaged within the estimator. A kinetic Monte Carlo simulator based on an 80×80 lattice model is used to describe the evolution of the process. The time interval between two available measurements is taken to be $0.3\,\text{s}$, which is consistent with available techniques that can be used to measure the growth rate in real time [121]. The substrate temperature is kept constant at $800\,\mathrm{K}$, and the initial inlet precursor mole fraction is 2.0×10^{-5}; these values correspond to a growth rate of about $180\,\mathrm{ML/s}$. The desired set-point value for the growth rate is $220\,\mathrm{ML/s}$. The parameters for the growth rate estimator and the PI controller used in this simulation are shown in Table 3.2 (column MIMO: Growth Rate). Figure 3.18 shows the growth rate and the inlet precursor mole fraction under feedback control. The results demonstrate that the growth rate can be successfully controlled to the desired set point by manipulating the precursor mole fraction.

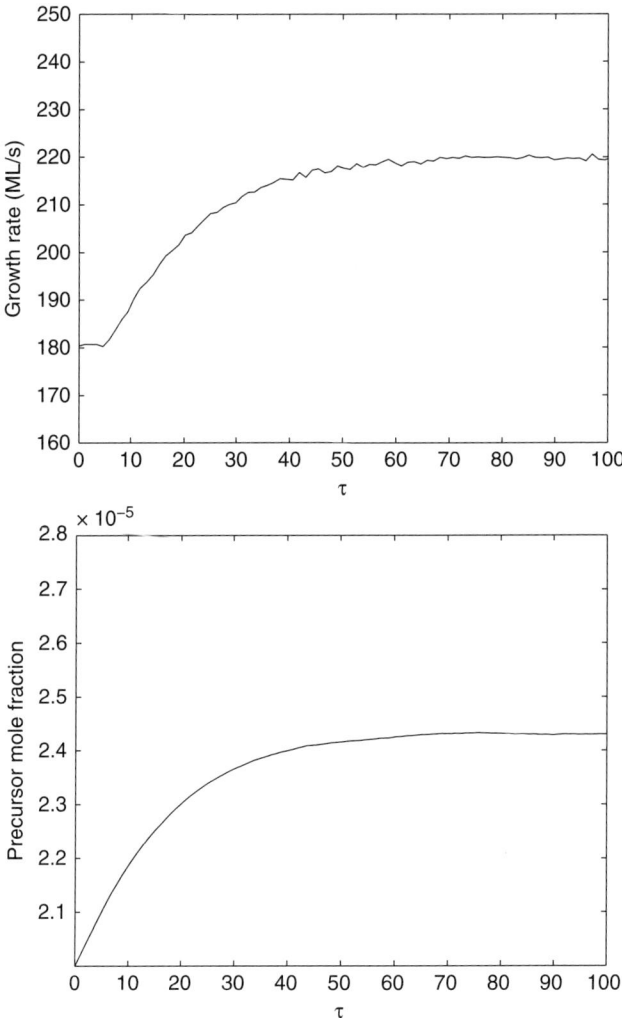

Fig. 3.18. Profiles of the growth rate and the inlet precursor mole fraction under single-loop feedback control using the estimator/controller structure of Fig. 3.6.

Table 3.2. All estimator and controller parameters.

SISO		MIMO			
roughness		Growth rate		Roughness	
K_s	1.00	K_s	1.00	K_s	1.00
K_e	0.08	K_e	0.08	K_e	0.08
K_c	30.00	K_c	2.0×10^{-9}	K_c	-15
τ_c	0.40	τ_c	0.40	τ_c	0.30
K_0	0.50	K_0	0.50	K_0	0.50
y_{set}	1.50				

In the second simulation run, the surface roughness is controlled by manipulating the substrate temperature while the inlet precursor mole fraction is kept constant. Figure 3.19 shows the profiles of the surface roughness and substrate temperature in the closed-loop system. A kinetic Monte Carlo simulator that uses an 80×80 lattice model is used to describe the evolution of the process. The inlet precursor mole fraction is kept constant at 2.0×10^{-5} and the substrate temperature is initially 800 K. The desired set-point value of surface roughness is 1.5. The parameters for the roughness estimator and controller are shown in Table 3.2 (column MIMO: Roughness). The controller successfully drives the surface roughness to the set point by manipulating the substrate temperature.

Effect of Multivariable Input–Output Interactions

The objective of this subsection is to understand the influence of multivariable input–output interactions on closed-loop performance and determine whether there is a need for the design and implementation of a multivariable controller that compensates for the effect of such interactions. To this end, we consider the problem of simultaneous regulation of the growth rate and of the surface roughness and use a feedback control system that is comprised of the estimator and two single-loop proportional-integral controllers. Specifically, following up on the results of the previous subsection, the input–ouput pairs are substrate temperature (T)–surface roughness (r) and inlet precursor mole fraction (y)–growth rate (gr).

A closed-loop system simulation is carried out to evaluate this approach to simultaneously control the growth rate and surface roughness. In this simulation, the outputs from six kinetic Monte Carlo simulators running on 20×20 lattice models are averaged within the estimator. A kMC simulator that uses an 80×80 lattice model is used to describe the evolution of the process. The parameters for the estimator and the controllers are listed in Table 3.2 (column MIMO). Compared to the results shown in Figs. 3.18 and 3.19, we find very similar roughness profiles (the reader may refer to [98] for detailed simulation results). However, the transient response of the growth rate in the case of simultaneous growth rate and surface roughness control is slower compared to the case where the growth rate is the only controlled output and the structure of Fig. 3.6 is used (Fig. 3.20). The reason for the slower transient response of the growth rate is the effect of multivariable input–ouput interactions

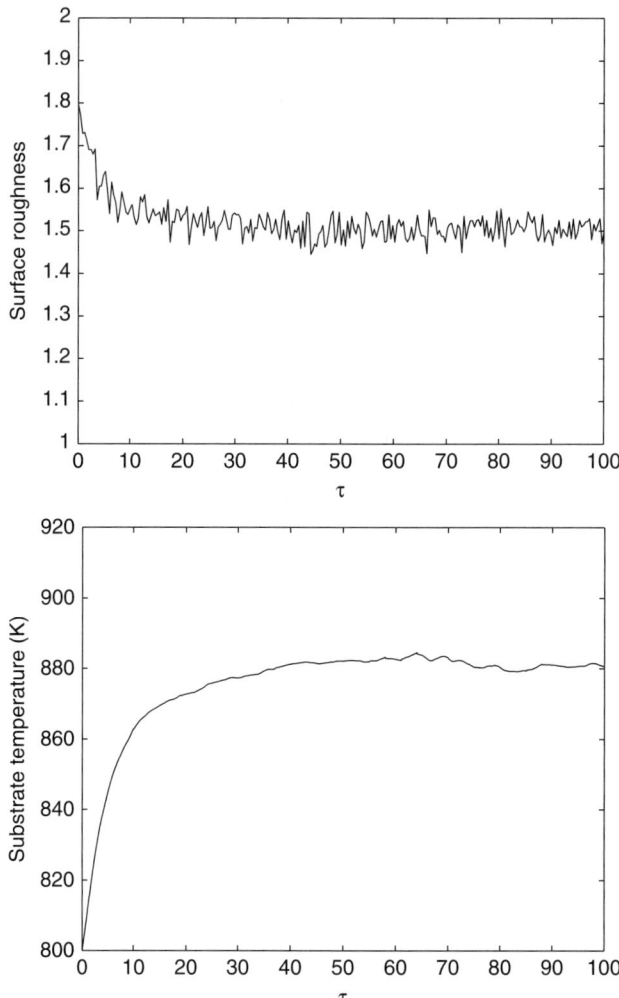

Fig. 3.19. Profiles of the surface roughness and the substrate temperature under single-loop feedback control using the estimator/controller structure of Fig. 3.6.

(i.e., the influence of the variation of substrate temperature on the growth rate and the influence of the variation of inlet precursor mole fraction on surface roughness in the closed-loop system); these interactions need to be compensated for in order to speed up the growth rate response and improve the closed-loop performance.

Multivariable Feedback Control of Surface Roughness and Growth Rate

This subsection focuses on the design and evaluation of a multivariable feedback control structure. To this end, we need to understand and model the interactions

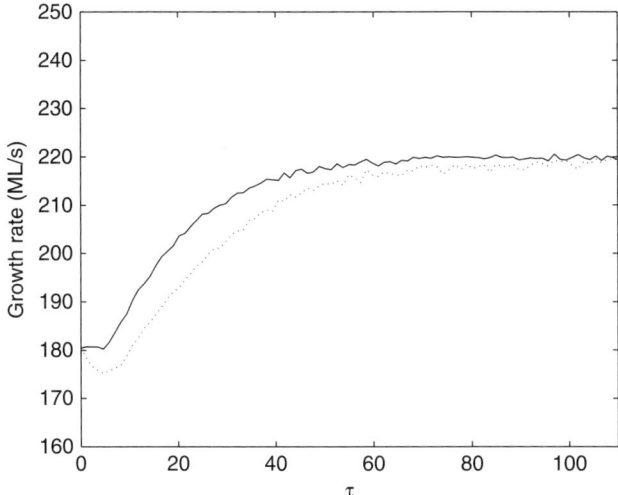

Fig. 3.20. Comparison of closed-loop growth rate profiles: (a) Growth rate is the only controlled output and the structure of Fig. 3.6 is used (solid line); (b) Simultaneous regulation of growth rate and surface roughness is considered and feedback control that does not account for multivariable input–output interactions is used (dotted line).

between inputs and outputs; two simulations are carried out to observe the changes in the surface roughness and growth rate for (1) a step change on substrate temperature with constant inlet precursor mole fraction, and (2) a step change on inlet precursor mole fractions with constant substrate temperature.

Figure 3.21 shows the growth rate and surface roughness when the substrate temperature is kept at $800\,\mathrm{K}$ and the inlet precursor mole fraction changes from 2.0×10^{-5} to 2.1×10^{-5} at $\tau = 10$. The results show that the growth rate increases from around $180\,\mathrm{ML/s}$ to $190\,\mathrm{ML/s}$, but there is no observable change in the surface roughness, which means the change in the inlet precursor mole fraction has very little influence on the surface roughness.

The interactions between the substrate temperature and the surface roughness and growth rate are studied by keeping the inlet precursor mole fraction at 2.0×10^{-5} and increasing the substrate temperature from $800\,\mathrm{K}$ to $840\,\mathrm{K}$ at $\tau = 10$. Figure 3.22 shows the responses of the growth rate and surface roughness to the substrate temperature change. The simulation results show that the growth rate drops from approximately $180\,\mathrm{ML/s}$ to $175\,\mathrm{ML/s}$ and the roughness drops from approximately 1.8 to 1.6, which indicates that variations on the substrate temperature influence both the surface roughness and growth rate. It should be pointed out that this specific coupling pattern is valid only under the specific growth conditions used in this simulation.

To improve the closed-loop performance, a multivariable feedback control structure is developed, that explicitly compensates for the effect of the multivariable interactions occurring in the process. The controller structure is obtained by introducing a compensation block between the multiple single-loop controllers and the process.

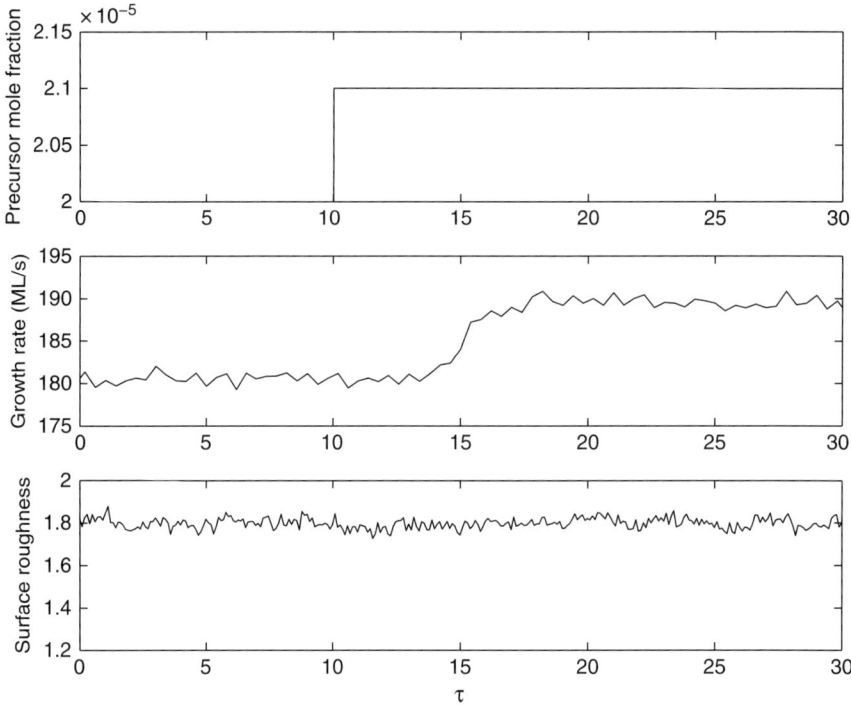

Fig. 3.21. Growth rate (middle plot) and surface roughness (bottom plot) profiles for a step change in inlet precursor mole fraction (top plot) from 2.0×10^{-5} to 2.1×10^{-5}. The substrate temperature is kept at $800\,\mathrm{K}$.

Because the interactions between the inlet precursor mole fraction and surface roughness are not significant (as shown in Fig. 3.21, bottom plot), only one compensator is needed in this example. A diagram of the multivariable control system using the estimator/controller structure with interaction compensation is shown in Fig. 3.7. $G_1(s)$ is the transfer function between the substrate temperature and the growth rate, and $G_2(s)$ is the transfer function between the inlet precursor mole fraction and the growth rate. Step tests are used to identify the expression and parameters of $G_1(s)$ and $G_2(s)$. Specifically, based on the simulation results shown in Fig. 3.21 (middle plot) and Fig. 3.22 (middle plot), G_1 and G_2 are taken to be constants with the following values: $G_1(s) = 0.125$ and $G_2(s) = 1.0 \times 10^7$. The rest of the process and controller parameters used in the simulation are the same as those in Table 3.2 (column MIMO).

A closed-loop system simulation is performed to evaluate the effectiveness of the multivariable estimator/control structure with interaction compensation shown in Fig. 3.7. The size of the small lattice is 20×20 and the outputs from six kinetic Monte Carlo simulators based on 20×20 lattice models are averaged within the estimator. A kinetic Monte Carlo simulator based on an 80×80 lattice model is

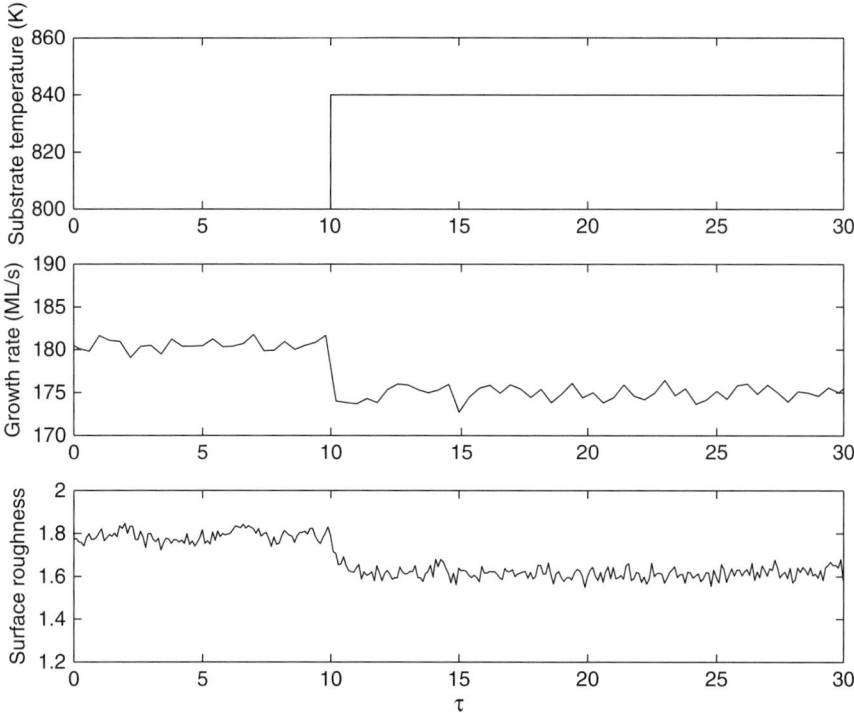

Fig. 3.22. Growth rate (middle plot) and surface roughness (bottom plot) with a step change in substrate temperature (top plot) from $800\,\mathrm{K}$ to $840\,\mathrm{K}$. The inlet precursor mole fraction is 2.0×10^{-5}.

used to describe the evolution of the process. The roughness set-point value is 1.5 and the growth rate set-point value is $220\,\mathrm{ML/s}$. Initially, the substrate temperature is $T = 800\,\mathrm{K}$ and the inlet precursor mole fraction is 2.0×10^{-5}; these conditions correspond to a growth rate of about $180\,\mathrm{ML/s}$ and a surface roughness of about 1.8. The proposed multivariable control system (Fig. 3.7) is applied to the process to regulate the growth rate and surface roughness to the desired set-point values. The left plot of Fig. 3.23 shows the comparison of the growth rate profiles under multivariable feedback control with interaction compensation and under multiple single-loop control without interaction compensation. By using the interaction compensator, the growth rate converges to the desired set-point value faster. The right plot of Fig. 3.23 shows the surface roughness under multivariable control with interaction compensation. The microstructure of the thin film at the beginning and at the end of the closed-loop system simulation run is shown in Fig. 3.24. These results show that the proposed multivariable control system with interaction compensation can simultaneously drive the growth rate and surface roughness to the desired set-point values and improve the closed-loop response.

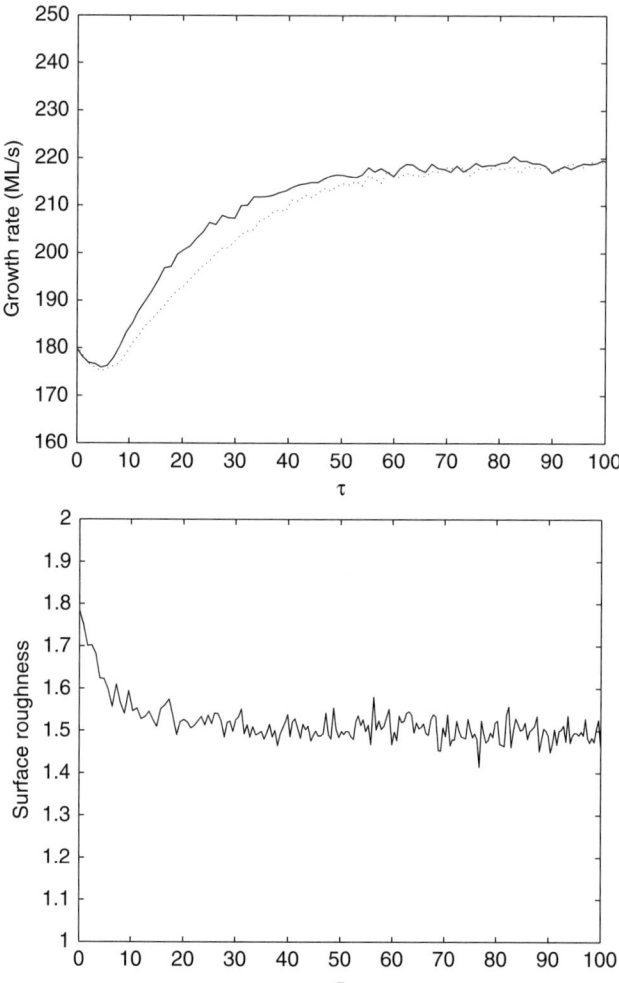

Fig. 3.23. Comparison of the closed-loop growth rate under multivariable feedback control with interaction compensation (solid line) and under multiple single-loop control (dashed line) (top plot) and closed-loop surface roughness under multivariable feedback control with interaction compensation (bottom plot).

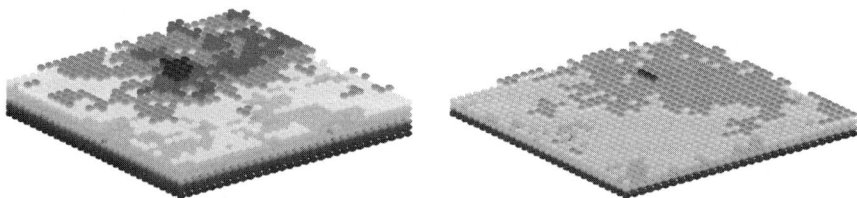

Fig. 3.24. Surface microconfiguration at the beginning (left plot, roughness $= 1.8$) and at the end of the closed-loop simulation run (right plot, roughness $= 1.5$).

3.5 Application to a Complex Deposition Process

This section investigates a complex deposition process, which includes two types of macromolecules whose growth behaviors are very different. This deposition process is influenced by both short-range and long-range interactions. The study of this process is motivated by recent experimental results on the growth of high-κ dielectric thin films using plasma-enhanced chemical vapor deposition (PECVD). An example is the PECVD ZrO_2 process [25], in which two major species, zirconium hydroxide and hydrocarbon, are present in the gas phase during the deposition. Recent experimental results have shown that when zirconium hydroxides are the dominant species in the gas phase, the deposited ZrO_2 thin film has a very smooth surface with a roughness value of less than half ZrO_2 monolayer. On the other hand, when hydrocarbons dominate the gas phase, the deposited ZrO_2 thin film has a very rough surface characterized by big islands, which suggests that the aggregation of the hydrocarbon species on the substrate surface, as a result of long-range interactions, is quite significant [24].

These results imply that a single-component kMC model considering only short-range interactions is inadequate to describe the thin-film growth in such a complex process. Therefore, a computationally efficient kMC model of heterogeneous deposition processes in which long-range interactions are accounted for is needed. Motivated by this, a multicomponent kMC model is developed for the deposition process. Both single-component and multicomponent cases are simulated and the dependence of the surface microstructure of the thin film, such as island size and surface roughness, on substrate temperature and gas-phase composition is studied. The surface morphology is found to be strongly influenced by these two factors, and growth regimes governed by short-range and long-range interactions are observed. Furthermore, two kMC model-based feedback control schemes that use the substrate temperature to control the final surface roughness of the thin film are proposed. The closed-loop simulation results demonstrate that robust deposition with controlled thin-film surface roughness can be achieved under a kMC estimator-based PI feedback controller in the short-range interaction-dominated growth regime, while a kMC model-predictive controller is needed to control the surface roughness in the long-range interaction-dominated growth regime.

3.5.1 Process Description

Deposition processes such as PECVD often involve large numbers of participating species with heterogeneous growth behaviors. Here, we study a heterogeneous deposition process in which two types of macromolecules of very different growth behavior, type A and type B, are present. A type A macromolecule is significantly affected by long-range attractions and tends to aggregate with other A macromolecules into clusters, i.e., it favors Volmer–Weber (VW) growth mode [61]. Hydrocarbon molecules generated from the decomposition of metallo-organic (MO) precursors in a PECVD process are good examples of such a type. A type B macromolecule favors surface sites of a local minimum height, which usually results in Frank–van der

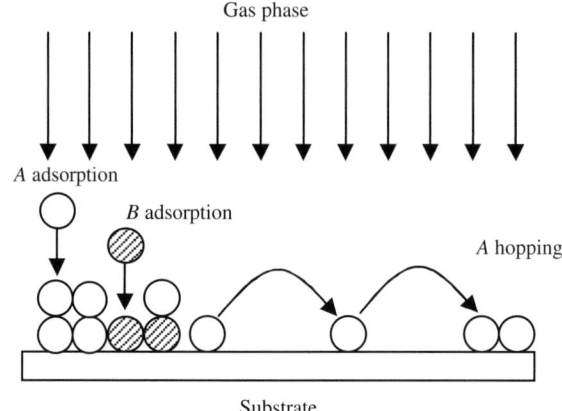

Fig. 3.25. The complex heterogeneous thin-film deposition process.

Merwe (FM) type of film growth [61]. Metal oxides or hydroxides originated from the MO precursors may behave similarly to type B macromolecules as discussed in the introduction.

The geometry of the deposition process is shown in Fig. 3.25. The gas flux is perpendicular to the substrate surface. Flux compositions, i.e., the flux of A and B, in terms of the number of macromolecules encountered per unit time per surface site, are taken as macroscopic process parameters. They can be measured directly (via mass spectrometer for example) or determined based on the measurements of the partial pressures for each species and gas-phase temperature using kinetic theory [90]. Thus, we model only the microprocesses taking place on the substrate surface. Both A and B can diffuse from the gas phase onto the substrate; however, type B macromolecules settle to surface sites of local minimum height (surface relaxation) simultaneously during adsorption. Surface migration and desorption processes are ignored (generally true for low-temperature CVD processes such as PECVD), while hopping of type A macromolecules is allowed (as if A is first physisorbed). Surface reactions are not explicitly considered in this process; however, the long-range behavior of A and the surface relaxation of B could be consequences of surface reactions (i.e., surface-mediated).

Surface Microstructure Model

The surface microstructure model is constructed based on a standard kMC scheme [57] that assumes the growth process to be a Poisson process. Therefore, the dynamics of the deposition process are governed by the master equation that describes the evolution of probabilities of the surface being in specific micro-configurations. Monte Carlo simulation is used to obtain realizations of this stochastic process that are consistent with the master equation.

To simulate the heterogeneous deposition process studied in this chapter, a simple cubic lattice structure, which is a good approximation for amorphous films, is used. The simulated surface domain is a square grid of 100 lattice points by 100 lattice points. To improve computational efficiency, the solid-on-solid assumption is made (i.e., voids and overhangs can be neglected). We consider a multilayer growth and assume that all the sites are available for adsorption of all gas-phase species at all times, and, thus, the adsorption rates of A (w_a^A) and B (w_a^B) are taken to be site-independent.

To incorporate different growth behaviors into a kMC scheme, a viable way is to set up specific rules for the microprocesses considered in the scheme (see [132] for an example of rule-based modeling of a coating microstructure). Although such rules may be arbitrary and may sacrifice the model's fidelity with respect to the detailed physics and chemistry, they are very favorable from a computational point of view and thus are preferable for real-time applications. Furthermore, when mechanisms of such behaviors are unknown, which is true for most of the complex PECVD processes, rule-based modeling must be used. In this work, we set up two rules for aggregation of the type A macromolecule and surface relaxation of the type B macromolecule, respectively.

For type A macromolecules, we enforce a rule on the hopping process. To encourage the aggregation of the surface A macromolecules over a longrange, we select the hopping direction of a type A macromolecule based on the number of the surface As that the hopping A can see in each direction instead of randomly picking among the possible hopping directions. Specifically, the hopping direction of a specific A macromolecule is determined by comparing the distance-weighted sum of all the A macromolecules in each direction and picking the largest sum. The weighted sum, for example, in the positive x-direction of an A located at the surface lattice point (x_0, y_0), $N_{h,+x}^A(x_0, y_0)$, is computed as follows:

$$N_{h,+x}^A(x_0, y_0) = \sum_{x=1}^{l_a} \sum_{y=-x}^{x} S_{(x_0+x, y_0+y)}^A \left(1 - \frac{\sqrt{x^2 + y^2}}{l_a} \right), \qquad (3.10)$$

where l_a is the maximum range of attraction, and the value of the occupancy factor $S_{(i,j)}^A$ is unity when the surface site (i, j) is occupied by an A and zero otherwise. Equation (3.10) imitates the sight of a surface A that fades out with distance, i.e., the near neighbors are weighted more than the distant neighbors. $1 - (\sqrt{x^2 + y^2}/l_a)$ is picked as the weighting function to employ a linearly decaying weighting that goes to zero at the boundary of the attraction zone. However, one can use any meaningful weighting function here to carry out simulations to simulate or validate specific deposition mechanisms.

The rate of the surface hopping of a type A macromolecule depends on the local activation energy barrier. Considering only the interactions of the first nearest side neighbors and the first nearest bottom neighbor to determine the hopping rate at a specific site, the hopping rate of a type A macromolecule on the surface with n first nearest side neighbors is given by

$$w_h^A(n) = k_{h0}^A \exp\left(-\frac{E_s^A + nE_n^A}{kT}\right),\tag{3.11}$$

where k_{h0}^A is the hopping frequency constant, and E_s^A and E_n^A are the energy barriers associated with the surface hopping of A for the bottom and side neighbors, respectively (we note that for simplicity we do not distinguish the neighboring macromolecules of different types).

For type B macromolecules, we enforce a surface relaxation rule on the adsorption process. During an adsorption event, a site (i, j) is first randomly picked among the sites of the whole lattice; the final site onto which the B macromolecule adsorbs could be different from the initially chosen site for the adsorption event. In particular, when the initially chosen site does not have the local minimum height, B will be adsorbed onto one of the neighboring sites that has the local minimum height. In this work, only the four first-nearest neighbor sites and four second-nearest neighbor sites are considered. In addition, the sticking probability of the type B macromolecule on the surface site occupied by type A macromolecules is considered very small (5% in this study). This is because when this sticking probability is close to unity, the surface would be smoothed by type B macromolecules independently of the presence of type A macromolecules; thus, the dynamics of the two-component deposition would not be observable. All other sticking probabilities are considered to be unity for simplicity.

The lifetime of each Monte Carlo event in the simulation τ can be determined by the following expression (see Section 2.4.2 for a detailed proof):

$$\tau = -\frac{\ln \xi}{w_a^A + w_a^B + \sum_{n=0}^{4} N_n^A w_h^A(n)},\tag{3.12}$$

where ξ is a random number that follows the uniform distribution in the unit interval and N_n^A is the number of type A macromolecules on the surface with n first-nearest side neighbors.

Remark 3.2. Since we treat hopping events toward different hopping directions as different microscopic events, the kinetic Monte Carlo simulation does not need to determine in which hopping direction the macromolecule needs to go once the event is selected; therefore, overriding the direction chosen by kMC by the microprocess rule is not an issue here, and the time increment can be calculated using Eq. (3.12). Hopping processes toward different hopping directions are treated as independent Poisson processes just like the adsorption of A and the adsorption of B. As long as the dynamical hierarchy of transition rates is preserved in the kinetic Monte Carlo simulation, the time increment should be selected from the exponential distribution of Eq. (3.12).

Simulation Procedure

The parameters, k_{h0}^A, E_s^A, E_n^A, and l_a in the model can be determined by optimal parameter estimation using experimental data (see also Section 4.4). However, the

Table 3.3. Parameters of the complex deposition process.

Hopping freq. const.	k_{h0}^A	10^{13}	s^{-1}
Hopping energy (bottom)	E_s^A	0.8	eV
Hopping energy (side)	E_n^A	0.2	eV
Attraction range	l_a	20	units

parameters used in this study are typical but arbitrarily chosen, and are shown in Table 3.3. When the lattice is set and the rates of the three events (A adsorption, B adsorption, A hopping) are determined based on measurements or its corresponding rate expression [Eq. (3.11)], a kinetic Monte Carlo simulation is executed following the algorithm in Section 2.4.2. First, the surface A macromolecules are grouped into five classes based on the number of side neighbors (from zero to four side neighbors); in each class, the macromolecules have the same hopping rates. However, they may have different hopping directions depending on the surface microconfiguration; the adsorption rates of the A and B macromolecules are both site-independent. Then, a random number is generated to select an event to be run based on the rates; if the event is A hopping, the class in which the event will happen is also selected. After that, a second random number is generated to select the site where the event will be executed; if the event is A or B adsorption, the site is randomly picked from sites in the entire lattice; if the event is A hopping, the site is randomly picked from the list of the sites in the selected class. After the site is selected, the MC event is executed. If the event is adsorption, it is executed by adding one macromolecule to the selected site (B adsorption rule is applied if the event is B adsorption); if the event is A hopping, the type A macromolecule on the site is moved to the next site in the direction selected by the hopping rule. Upon an executed event, a time increment τ computed based on Eq. (3.12) is added to the process time t. Periodic boundary conditions are used in the simulation to satisfy the mass balance of the hopping macromolecules.

3.5.2 Open-Loop Dynamics

Using the proposed growth model, a parametric analysis of the growth process is conducted. Specifically, we study the effects of the substrate temperature (in the range of 300 K to 440 K, which is the normal operating temperature of low-temperature CVD processes such as PECVD) and gas-phase composition on the surface microstructure of the deposited thin films in both homogeneous and heterogeneous deposition processes. The effect of the simulation lattice size is also investigated. Furthermore, island sizes of the thin films obtained under different process conditions are qualitatively compared and the surface roughness of each film, r, is computed in a root-mean-square fashion using Eq. (3.2). This study provides valuable insight for the formulation of the control problem. In this section, we summarize the open-loop dynamics of the complex deposition process. The reader may refer to [116] for detailed simulation results.

Single-Component Case

The simulation of a single-component deposition process can be executed by setting either parameter w_a^A or w_a^B in the heterogeneous model equal to 0.

It has been shown that the surface of the thin film obtained by a deposition with only type B macromolecules is very smooth due to the surface relaxation of B, and the film growth is in Frank–van der Merwe mode. Since the adsorption rate is independent of the substrate temperature, the surface microstructure of the thin film has no dependence on the substrate temperature.

On the other hand, the surface morphologies of thin films obtained by depositions with only type A macromolecules present in the gas phase strongly depend on the substrate temperature. Specifically, the thin film deposited at a low substrate temperature ($T = 320$ K) has a high island density but a small lateral island size. The thin film deposited at a high substrate temperature ($T = 380$ K) has a low island density but a large lateral island size.

The very different surface microstructure observed for these two films, which have similar roughness values, can be explained by the different growth modes in these two temperature regimes. According to the hopping rate equation [Eq. (3.11)], surface hopping of type A macromolecules has an Arrhenius-type dependence on the substrate temperature T. Thus, at a low substrate temperature, the hopping rate is much smaller than the rate at a high substrate temperature. This suggests that at low temperature, the dominant surface microprocess is the adsorption process. Therefore, although the long-range attraction tends to drive the surface type A macromolecules together, the hopping rate is so low that these macromolecules are not able to move along the direction of attraction far enough to form large islands. Therefore, the effect of long-range attraction is not significant and the aggregation mostly occurs in the vertical direction (one-dimensional aggregation) by A adsorption. This growth mechanism results in a surface with islands of large height and small lateral size.

On the other hand, when the substrate temperature is high, the rate of hopping becomes large, and surface type A macromolecules are able to move along the direction of attraction for a distance comparable to the range of attraction. Therefore, the effect of long-range attraction becomes very significant and aggregation occurs in both vertical and horizontal directions (three-dimensional aggregation, i.e., VW growth mode) by adsorption and hopping. This growth mechanism leads to the formation of islands with large dimensions in both the vertical and horizontal directions. Furthermore, we note that although the range of attraction is limited to l_a, the lateral size of the islands is not limited by the range of interaction due to the coalescence between islands. Islands of lateral size larger than l_a are observed in our simulations for high-substrate-temperature depositions.

Time evolution of the surface roughness is also investigated through kMC simulations. Our simulation results show that the thin-film growth in the B-only deposition is 2D growth (the surface roughness saturates over time) while the growth in the A-only deposition, with either low or high substrate temperature, is 3D growth (the surface roughness never saturates).

Multicomponent Case

We have also simulated a heterogeneous multicomponent deposition using the proposed process model. Both types A and B macromolecules are present in the gas phase, and the relative ratio of the two species is set to be unity in the simulated case for simplicity.

Figure 3.26 shows the surface morphology of thin films obtained by depositions at low ($T = 320$ K) and high ($T = 440$ K) substrate temperatures, respectively. The difference in surface morphology between the two thin films is similar to the single-component case in which only type A macromolecules are present in the gas phase. This is expected since the behavior of the two types of macromolecules is considered independent of each other in the simulation.

Figure 3.27 shows the surface roughness of thin films deposited at different substrate temperatures. It can be seen that there are two temperature regimes in which thin-film growth is quite different. In the low-temperature regime, the surface roughness drops with increasing temperature, while in the high-temperature regime, the surface roughness rises with increasing temperature (however, the surface roughness drops again when the substrate temperature is very high when stable surface islands start to coalesce and form islands with lateral dimension larger than the range of attraction). Based on the discussion above about the different growth modes at low and high substrate temperatures, the transition from the low-temperature regime to the high-temperature regime corresponds to the change in growth process from short-range interaction-dominant to long-range attraction-dominant. The substrate temperature at which the minimum roughness is achieved corresponds to the separation of the two temperature regimes.

Furthermore, since different gas-phase compositions can be used to tailor the material properties of the thin films obtained by the depositions for different applications, simulations of depositions for different gas-phase compositions have been run to study the effect of gas-phase composition on the surface microstructure based on the proposed growth model. It can be observed that the increase in the relative concentration of A leads to increasing surface roughness, and vice versa. Moreover, when the relative concentration of A is larger than 50%, the presence of two temperature regimes is quite clear. This again suggests that type A macromolecules have a significant effect on the surface microstructure when the relative concentrations of A and B are comparable [116].

Effect of Lattice Size

Because the dimension of the wafers used in a real deposition process is usually in the 10^8 nm regime, it is impossible to simulate the film growth for the entire wafer even with the most efficient Monte Carlo algorithm and the best available computing power. Thus, a lattice size that corresponds to a very small spatial domain compared to the actual wafer dimension is used in this work. However, to better gauge the results from such simulations, investigating the lattice-size dependency of the simulation results is necessary.

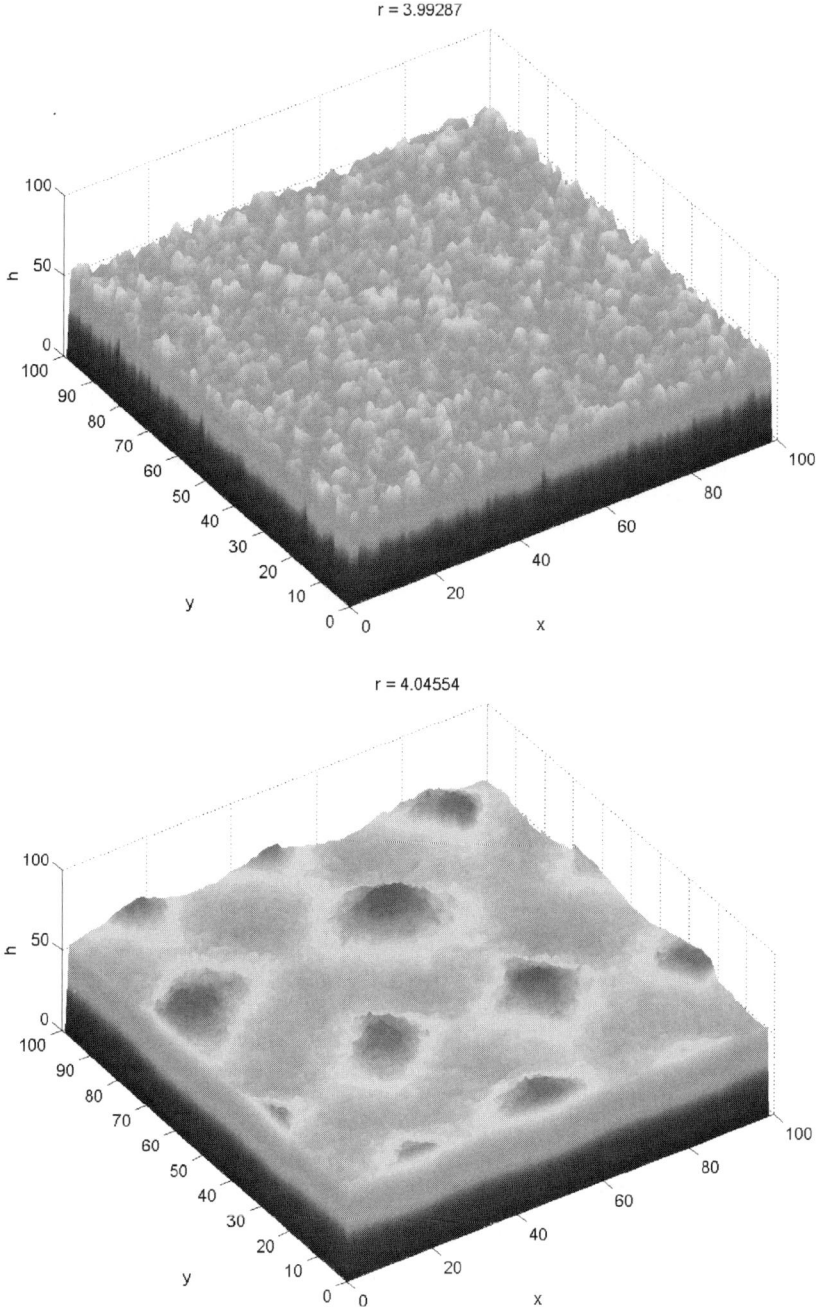

Fig. 3.26. Surface of a thin film deposited with $w_a^A = 0.05$ s^{-1}, $w_a^B = 0.05$ s^{-1} and different temperatures at $t = 900$ s: $T = 320$ K (left plot) and $T = 380$ K (right plot).

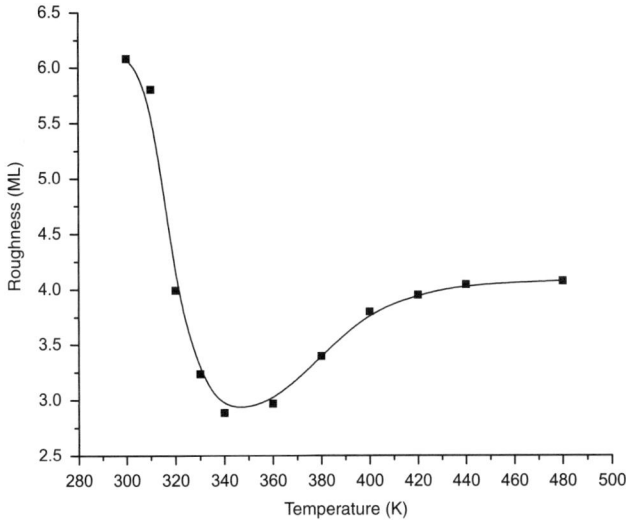

Fig. 3.27. Surface roughness of thin films deposited with $w_a^A = 0.05\,\mathrm{s}^{-1}$, $w_a^B = 0.05\,\mathrm{s}^{-1}$ for different substrate temperature. ($t = 900\,\mathrm{s}$).

Figure 3.28 shows the surface roughness of thin films deposited with different substrate temperatures computed using different simulation lattice sizes. It can be seen that simulation results from kMC runs with a lattice size larger than 50×50 agree very well with each other, while simulation results from the 50×50 lattice show qualitative agreement with the results from the larger lattice runs. However, the result from the 20×20 lattice is inconsistent with all other results. Such results suggest that the lattice size used in the kMC simulation should be at least twice as large as the surface interaction radius (20 in this work) to capture the dynamics of the thin-film growth process. Moreover, the stochastic noise of the simulation decreases with increasing lattice size. For this work in particular, a simulation lattice size of 100×100 is large enough to describe the process dynamics with low stochastic noise; therefore, such a lattice size (or larger) will be used in the subsequent simulations.

3.5.3 Low-Temperature Regime: PI Control Design

For thin-film applications where a high island density is desired, surface roughness control can be implemented on the low-temperature regime, in which the growth process is dominated by short-range interactions. Real-time feedback control of deposition processes that are characterized by short-range interactions has been discussed in detail in Section 3.4, where the control problem was formulated as regulation of instantaneous surface roughness. In this work, a feedback control scheme inspired by the methodology proposed in Sections 3.2 and 3.3 is developed to regulate the final surface roughness. To ensure that the surface of the thin film has a high island density, the substrate temperature is restricted within the range of 300 K–340 K.

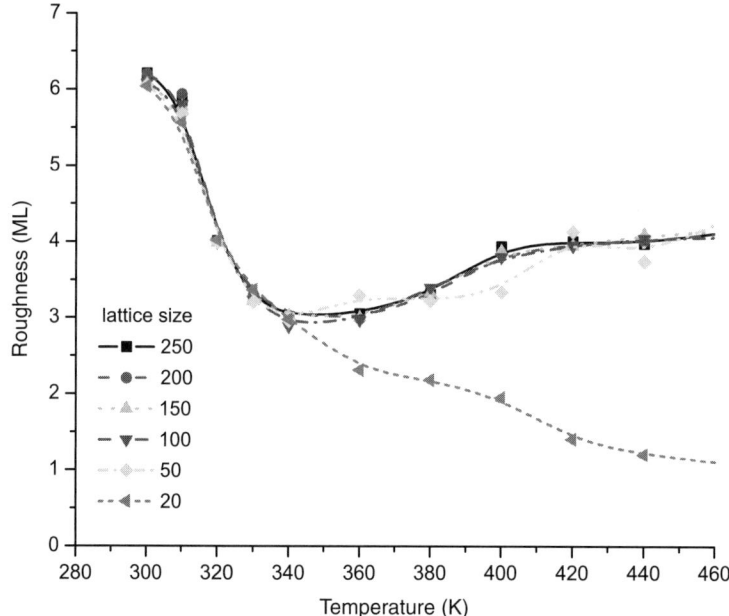

Fig. 3.28. Surface roughness of thin films deposited with different substrate temperatures computed using different simulation lattice sizes.

Open-Loop Response

We first investigate the open-loop response of surface roughness with respect to step changes in the substrate temperature and gas-phase composition. Our simulation results demonstrate that the value of the surface roughness at the end of the deposition can be controlled by manipulating the substrate temperature and the final surface roughness can be computed based on the current thin-film surface, the gas flux composition, and the substrate temperature using the proposed kMC growth model. Furthermore, it can be observed that increasing concentration of type A macromolecules in the gas phase, i.e., increasing w_a^A, results in a rise in the final thin-film surface roughness, and vice versa [116].

Controller Design: Closed-Loop Simulation

Based on the open-loop system analysis, a real-time surface roughness feedback control scheme is designed for the process. Figure 3.29 shows the block diagram of the closed-loop system. The thin-film growth is influenced by the substrate temperature and the macromolecule adsorption rates, which are determined by the gas-phase composition. Since we chose the final surface roughness as the variable to control, an estimate is computed for every control cycle using the proposed kMC model based on the operating conditions and the thin-film surface configuration, which

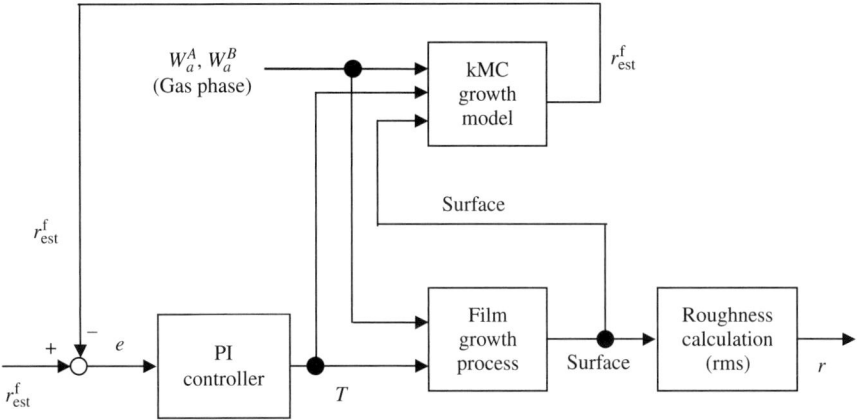

Fig. 3.29. Block diagram of the closed-loop system.

can be measured in real time by advanced surface characterization tools such as the GISAXS [129].

The substrate temperature is then computed based on a proportional-integral (PI) control algorithm as follows:

$$T(t) = K_c \hat{e}(t) + K_i \int_{t_0}^{t} \hat{e}(\mu) d\mu + T_0, \tag{3.13}$$

$$\hat{e}(t) = \begin{cases} e(t) & |e(t),| > \epsilon, \\ 0, & |e(t)| \leq \epsilon, \end{cases} \tag{3.14}$$

where $T(t)$ is the controller output (i.e., substrate temperature), T_0 is the initial substrate temperature, K_c is the proportional gain, K_i is the integral gain, $e(t)$ is the error of the final surface roughness (i.e., the difference between the desired final surface roughness and the estimated final surface roughness computed using the proposed kMC model based on the surface configuration and operating conditions at the instance of last control action), and ϵ is the error tolerance, which is used to improve the robustness properties of the controller against stochastic noises ($\epsilon = 0.05$ in this work).

The closed-loop thin-film growth process that employs the proposed real-time feedback control scheme has been simulated. The controller parameters K_c and K_i are set to be -0.1 and -0.5, respectively. For all the closed-loop simulations, the deposition duration is set to be $950\,\mathrm{s}$, and the adsorption rates w_a^A and w_a^B are both set to be $0.05\,\mathrm{s}^{-1}$. The estimator and controller are activated at $t = 400\,\mathrm{s}$, and the control action is applied to the process every $10\,\mathrm{s}$.

Figure 3.30 shows the temperature and surface roughness profiles with a final surface roughness set-point value of 3.5 ML. It can be seen that the surface roughness value of the thin film at the end of the deposition has been controlled at the desired value, which is 12.5% lower than the surface roughness of the thin film deposited by open-loop deposition with the same initial deposition conditions.

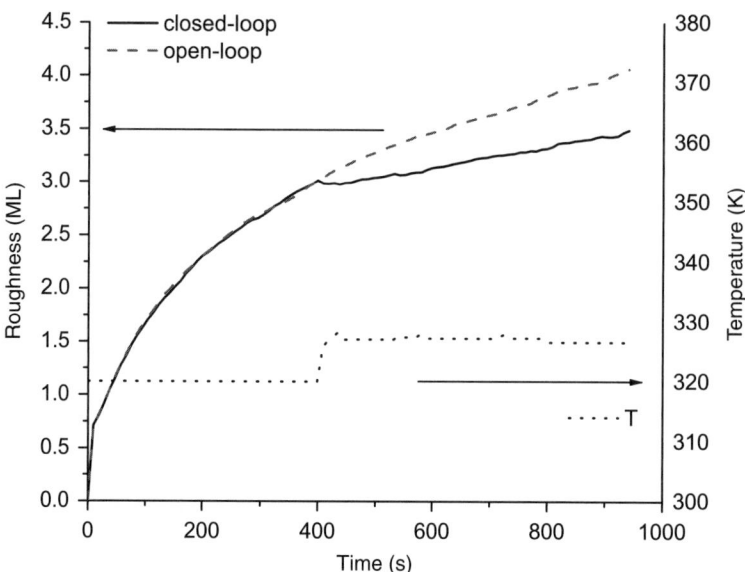

Fig. 3.30. Temperature and surface roughness profiles with surface roughness set-point value of 3.5 ML: (a) closed-loop surface roughness (solid line, left scale); (b) open-loop surface roughness (dashed line, left scale); (c) substrate temperature (dotted line, right scale).

Figure 3.31 shows the temperature and surface roughness profiles with final surface roughness set-point value of 3.5 ML. A disturbance in the gas-phase composition is introduced in this simulation represented by a step change in the adsorption rates at $t = 400$ s to $t = 500$ s. Specifically, w_a^A changed from $0.05\,\mathrm{s}^{-1}$ to $0.1\,\mathrm{s}^{-1}$ and w_a^B changed from $0.05\,\mathrm{s}^{-1}$ to $0\,\mathrm{s}^{-1}$ at $t = 400$ s, while at $t = 500$ s, w_a^A and w_a^B both changed back to $0.05\,\mathrm{s}^{-1}$. It can be seen that the surface roughness of the thin film at the end of the deposition has been controlled at the desired value in spite of the disturbance in the gas phase, while the thin film deposited by open-loop deposition has a surface roughness 19.2% higher than the desired value due to the disturbance.

3.5.4 High-Temperature Regime: MPC Design

For thin-film applications where a large island size is desired, surface roughness control can be implemented for the high-temperature regime, in which the growth process is dominated by long-range interactions. To ensure the surface of the thin film has a low island density, the substrate temperature is restricted within the range of 340 K–420 K.

Open-Loop Response

The open-loop response of surface roughness with respect to step changes in the substrate temperature and gas-phase composition at the high-temperature regime is also

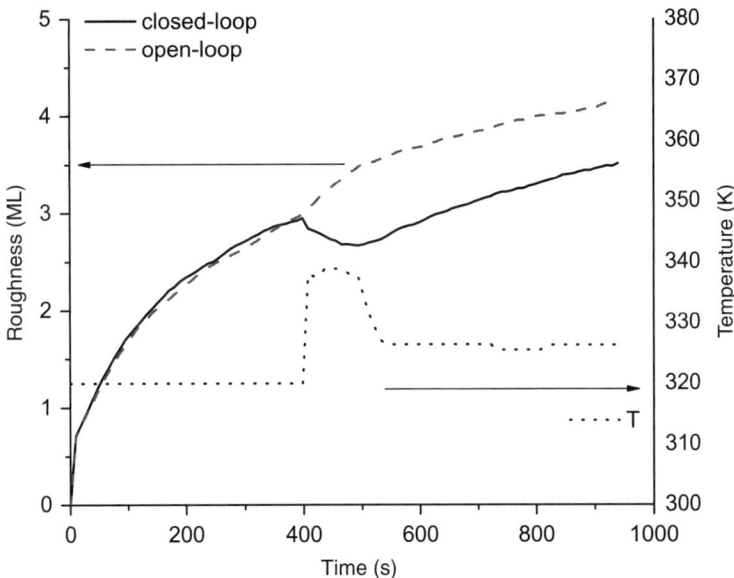

Fig. 3.31. Temperature and surface roughness profiles with surface roughness set-point value of 3.5 ML in the presence of disturbance: (a) closed-loop surface roughness (solid line, left scale); (b) open-loop surface roughness (dashed line, left scale); (c) substrate temperature (dotted line, right scale).

studied through kMC simulations. We can see that, in the high-temperature regime, the value of the surface roughness at the end of the deposition can also be controlled by manipulating the substrate temperature. However, both the responses to step changes in substrate temperature and gas-phase composition exhibit inverse dynamics, which suggests that a more advanced controller may be needed to control the surface roughness in the high-temperature regime [116].

Controller Design: PI Control

The high-temperature regime, closed-loop thin-film growth process that employs the proposed real-time feedback control scheme for the low-temperature regime but with different controller parameters has been simulated. The controller parameters K_c and K_i are set to be -2 and -0.2, respectively.

Figure 3.32 shows the temperature and surface roughness profiles with a final surface roughness set-point value of 3.2 ML. It can be seen that the surface roughness value of the thin film at the end of the deposition has been controlled at the desired value, which is about the same as the surface roughness of the thin film deposited by open-loop deposition with the same initial deposition conditions.

Figure 3.33 shows the temperature and surface roughness profiles with a final surface roughness set-point value of 3.2 ML. A disturbance in the gas-phase composition is introduced in this simulation in terms of a step change in the adsorption

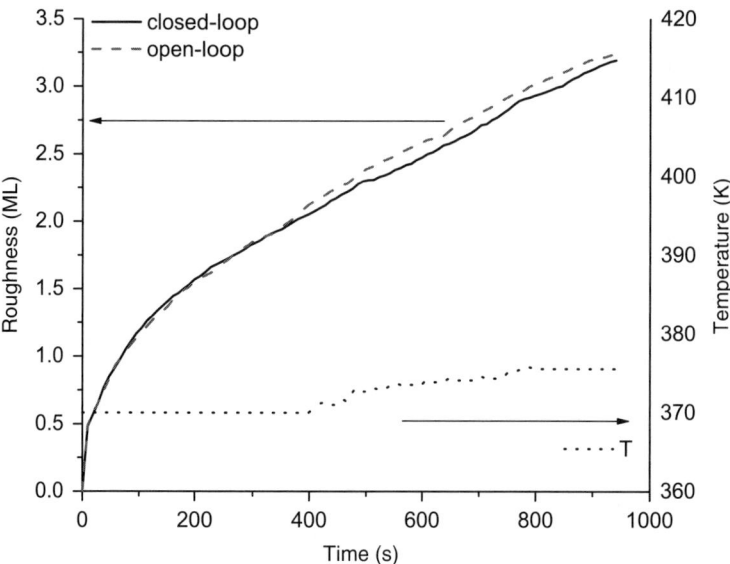

Fig. 3.32. Temperature and surface roughness profiles with a surface roughness set-point value of 3.2 ML: (a) closed-loop surface roughness (solid line, left scale); (b) open-loop surface roughness (dashed line, left scale); (c) substrate temperature (dotted line, right scale).

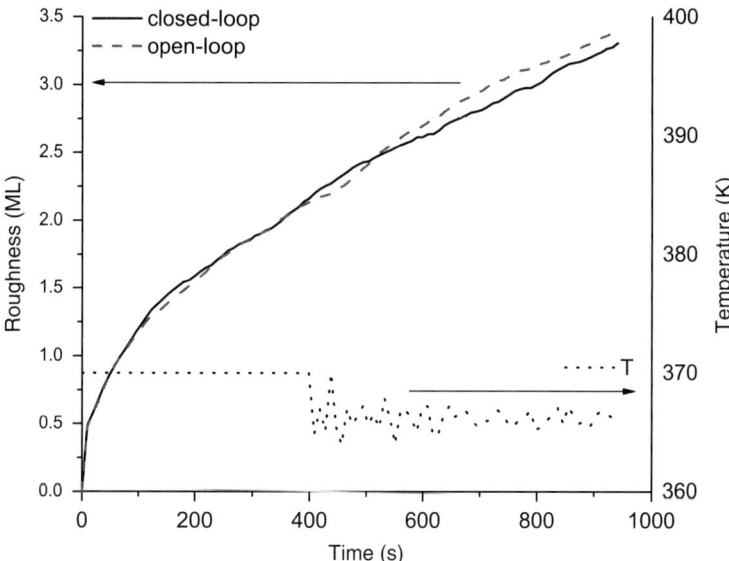

Fig. 3.33. Temperature and surface roughness profiles with a surface roughness set-point value of 3.2 ML in the presence of disturbance: (a) closed-loop surface roughness (solid line, left scale); (b) open-loop surface roughness (dashed line, left scale); (c) substrate temperature (dotted line, right scale).

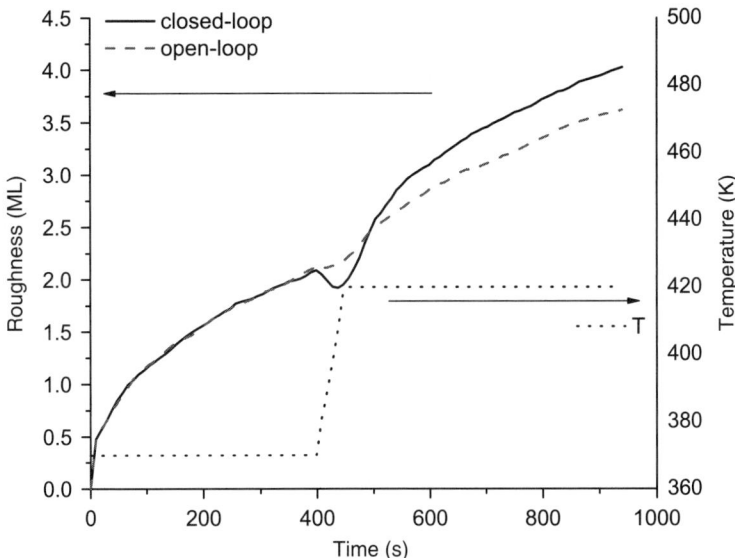

Fig. 3.34. Temperature and surface roughness profiles with a surface roughness set-point value of 3.2 ML in the presence of disturbance: (a) closed-loop surface roughness (solid line, left scale); (b) open-loop surface roughness (dashed line, left scale); (c) substrate temperature (dotted line, right scale).

rates at $t = 400$ s to $t = 420$ s. Specifically, w_a^A changed from $0.05\,\mathrm{s}^{-1}$ to $0.1\,\mathrm{s}^{-1}$ and w_a^B changed from $0.05\,\mathrm{s}^{-1}$ to $0\,\mathrm{s}^{-1}$ at $t = 400$ s, while at $t = 420$ s, w_a^A and w_a^B both changed back to $0.05\,\mathrm{s}^{-1}$. It can be seen that the final surface roughness has been controlled at the desired value, which is 5.9% lower than that obtained under open-loop operation in spite of the disturbance in the gas phase.

Figure 3.34 shows the temperature and surface roughness profiles with a final surface roughness set-point value of 3.2 ML. A disturbance in the gas-phase composition is introduced in this simulation in terms of a step change in the adsorption rates at $t = 400$ s to $t = 500$ s. Specifically, at $t = 400$ s, w_a^A changed from $0.05\,\mathrm{s}^{-1}$ to $0.1\,\mathrm{s}^{-1}$ and w_a^B changed from $0.05\,\mathrm{s}^{-1}$ to $0\,\mathrm{s}^{-1}$, while at $t = 500$ s, w_a^A and w_a^B both changed back to $0.05\,\mathrm{s}^{-1}$. It can be seen that the surface roughness is not controlled at the desired value and that the controller output hits the high limit of the control actuator. The final surface roughness is about 13.9% higher than that obtained under the open-loop operation. This suggests that the kMC estimator-based PI controller does not provide satisfactory closed-loop performance when the disturbance to the process is significant and, therefore, that a more advanced control scheme is needed.

Controller Design: kMC Model-Based Predictive Control

In order to achieve robust closed-loop operation in the high-temperature regime, a kMC model-based predictive control scheme is proposed. Figure 3.8 shows the block diagram of the closed-loop system. A reference trajectory of the instantaneous surface roughness of the thin film is selected based on offline optimization. For simplicity in this work, the profile of the surface roughness of the thin film in an ideal open-loop deposition (no disturbance is assumed to affect the process, and the final surface roughness is taken to be the desired value) is chosen. Using such a reference trajectory, instead of solving the receding-horizon optimization problem of minimizing the difference between the final surface roughness and the desired value with multiple decision variables, we need only to solve the fixed short-horizon optimization problem of minimizing the difference between the instantaneous surface roughness and the reference value with a single decision variable. Therefore, the computation time of each optimization is greatly reduced, since the kMC simulation duration is reduced from the scale of the total deposition time to the controller turnover time. This is very important since kMC simulation is relatively time-consuming and large-scale numerical optimization using kMC models is almost impossible to solve in real time (see also Chapters 6 and 7 for more on this topic).

During each control cycle, the surface configuration $X(k)$ (i.e., the height and the types of the top two macromolecules of each surface site) is first measured. An estimate of the surface configuration at the next control action time $X_{\mathrm{est}}(k+1)$ is computed based on the current process conditions using the proposed kMC model, and the estimated surface roughness value $r_{\mathrm{est}}(k+1)$ is compared with the reference value $r_{\mathrm{ref}}(k+1)$. If the error is less than ε ($\varepsilon = 0.05$ in this work), the next controller output $T(k+1)$ is set to be the same as the current output $T(k)$. If the error is larger than ε, the optimizer is called to compute the output value of the next control action $T(k+1)$ so that the error between the surface roughness after the next control action $r(k+2)$ and the reference value $r_{\mathrm{ref}}(k+2)$ is minimized.

The optimizer uses direct search to find the optimal solution since the kMC model does not have a closed-form expression. The estimate of the surface roughness after the next control action $r_{\mathrm{est}}(k+1)$ is computed using the proposed kMC model based on the estimated surface configuration before the next control action $X_{\mathrm{est}}(k+1)$, the probe output value $T_{\mathrm{probe}}(k+1)$, and current process conditions. The search precision specified in this work is 1 K, and since the proposed kMC model is highly computationally efficient, the optimization problem can be solved by an entry-level personal computer within the controller's turnover time (10 s). Furthermore, the speed and the precision of the direct search optimization algorithm can be substantially improved by parallel computing.

Figure 3.35 shows the temperature and surface roughness profiles with a final surface roughness set-point value of 3.2 ML. The reference trajectory is computed by averaging the open-loop surface roughness profiles from six independent simulation runs. It can be seen that the surface roughness value of the thin film follows the reference trajectory closely and the final surface roughness has been controlled at the desired value.

Figure 3.36 shows the temperature and surface roughness profiles with a final surface roughness set-point value of 3.2 ML. A disturbance in the gas-phase composition is introduced in this simulation in terms of a change in the adsorption rates at $t = 400\,\mathrm{s}$ to $t = 500\,\mathrm{s}$, specifically, w_a^A changed from $0.05\,\mathrm{s}^{-1}$ to $0.1\,\mathrm{s}^{-1}$ and w_a^B

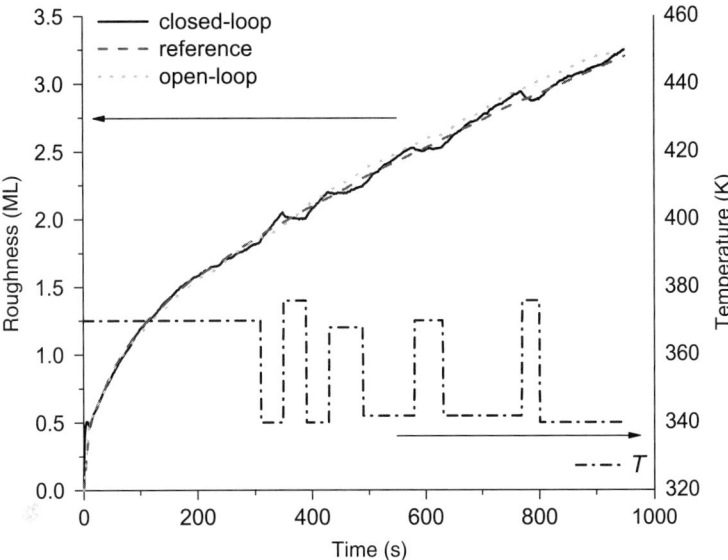

Fig. 3.35. Temperature and surface roughness profiles with a surface roughness set-point value of 3.2 ML: (a) closed-loop surface roughness (solid line, left scale); (b) reference surface roughness (dashed line, left scale); (c) open-loop surface roughness (dotted line, left scale); (d) substrate temperature (dashed dotted line, right scale).

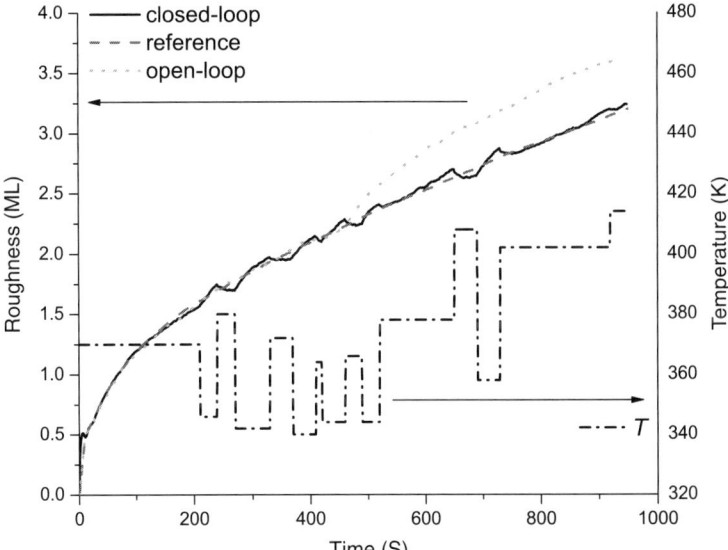

Fig. 3.36. Temperature and surface roughness profiles with a surface roughness set-point value of 3.2 ML: (a) closed-loop surface roughness (solid line, left scale); (b) reference surface roughness (dashed line, left scale); (c) open-loop surface roughness (dotted line, left scale); (d) substrate temperature (dashed dotted line, right scale).

changed from $0.05\,\mathrm{s}^{-1}$ to $0\,\mathrm{s}^{-1}$ at $t = 400\,\mathrm{s}$, while at $t = 500\,\mathrm{s}$, w_a^A and w_a^B both changed back to $0.05\,\mathrm{s}^{-1}$. It can be seen that the surface roughness closely follows the reference trajectory and that the final surface roughness has been controlled at the desired value, which is 13.5% lower than the open-loop value. Compared to the poor performance of the PI controller on the same closed-loop simulation case, the kMC model-based predictive controller delivers substantially improved and robust closed-loop performance.

3.6 Conclusions

This chapter presented methods for real-time estimation and control of surface roughness in thin-film growth. SISO, MIMO, and kMC model-predictive control systems, which can be implemented in real time, were developed and applied to thin-film deposition processes. The control systems use estimators that provide estimates of the controlled variables such as surface roughness and growth rate at a time scale comparable to the real-time evolution of the process. Specifically, an estimator/controller structure was initially developed and was applied to the multiscale process model of a thin-film growth process to regulate the surface roughness at a desired set-point value. Then, multivariable feedback control of surface roughness and growth rate in thin-film growth was studied and a multivariable feedback control system was proposed. Application of the multivariable feedback control system to the multiscale process model demonstrated successful regulation of both the surface roughness and growth rate to desired set-point values. Finally, two kMC model-based feedback control schemes were presented to control the final surface roughness of a complex deposition process, which included two types of macromolecules and both short-range and long-range interactions. The closed-loop simulation results demonstrated that robust deposition with controlled thin-film surface roughness could be achieved under a kMC estimator-based PI feedback controller in the short-range interaction-dominated growth regime, while a kMC model-predictive controller was needed to control the surface roughness in the long-range interaction-dominated growth regime.

4

Construction of Stochastic PDEs

4.1 Introduction

While it is possible in certain cases to use kinetic Monte Carlo models for real-time estimation and control of thin-film microstructure, there are many applications where closed-form models are needed, due to their computational efficiency, to carry out system-level analysis as well as design and implementation of real-time, model-based feedback control systems. For many deposition and sputtering processes, closed-form process models, in the form of linear or nonlinear stochastic partial differential equations (PDEs), are available (e.g., [36, 39, 91, 154, 158]). Stochastic PDEs contain the surface morphology information of thin films; thus, they may be used for the purpose of feedback controller design. For example, it has been experimentally verified that the Kardar–Parisi–Zhang (KPZ) equation [82] can describe the evolution of the surface morphology of gallium arsenide (GaAs) thin films, which is consistent with the surface measured by atomic force microscopy (AFM) [13, 79].

This has motivated recent research on the development of methods for feedback control of surface roughness based on linear and nonlinear stochastic PDE process models [102, 101, 117, 103]. In all the developments, a feedback controller designed based on a stochastic PDE process model can be applied to the kMC model of the same process to regulate the surface roughness to desired values.

However, the derivation of linear/nonlinear stochastic PDE models for complex thin-film growth processes directly based on microscopic process rules is a very difficult task. This issue has prohibited the development of stochastic PDE models, and subsequently the design of model-based feedback control systems, for realistic deposition processes that are, in general, highly complex. To address this issue, an initial effort was made to apply system identification techniques to identify parameters of a linear stochastic PDE model [101] using kinetic Monte Carlo simulation data. The results show that data-driven, optimization-based methods are appropriate to construct stochastic PDE models using data obtained from first-principles simulations. Inspired by this, methods for the construction of linear stochastic PDEs and

P.D. Christofides et al., *Control and Optimization of Multiscale Process Systems*,
Control Engineering, DOI 10.1007/978-0-8176-4793-3_4,
© Birkhäuser Boston, a part of Springer Science+Business Media, LLC 2009

parameter estimation of nonlinear stochastic PDEs were developed in a recent series of papers [118, 117, 74].

In this chapter, we present methods for the construction of linear and nonlinear stochastic PDEs. Linear stochastic PDEs on a one-dimensional (1D) spatial domain are first considered. A generic linear stochastic PDE model for thin-film deposition processes is initially reformulated into a system of infinite stochastic ordinary differential equations (ODEs) by using modal decomposition. The dependence of the statistical moments of the ODE states on the eigenvalues of the linear spatial operator of the stochastic PDE is subsequently derived. Then, we use a kMC simulation of the thin-film growth process to generate surface snapshots for different instants during process evolution to obtain values of the state vector of the stochastic ODE system. Using the kMC simulation data, a linear stochastic PDE model is determined by least-squares fitting of the prederivative coefficients to match the spectrum of the stochastic PDE system to the computed spectrum of the stochastic ODE system. The linear stochastic PDE construction method is then extended to two-dimensional (2D) applications. The method is applied to construct linear stochastic PDE models for thin-film growth processes taking place on 1D and 2D lattices.

Furthermore, the existence of nonlinearities in certain material preparation processes motivates the development of a method for the construction of nonlinear stochastic PDEs. Nonlinear stochastic PDE models are constructed by combining a priori knowledge on the model structure and a novel model parameter estimation procedure. To perform the parameter estimation, the nonlinear stochastic PDE is first formulated into a system of infinite nonlinear stochastic ODEs and then a finite-dimensional approximation is derived that captures the dominant mode contribution to the PDE state. The evolution of the state covariance of the stochastic ODE system is subsequently derived. Then, we use a kMC simulator to generate surface snapshots for different instants during process evolution to obtain values of the state vector of the stochastic ODE system. The model parameters of the nonlinear stochastic KSE are obtained by using least-squares fitting so that the state covariance computed from the stochastic KSE process model matches that computed from kMC simulations. The effectiveness of the method is demonstrated through application to an ion-sputtering process model described by the nonlinear stochastic Kuramoto–Sivashinsky equation (KSE). Some of the results of this chapter were first presented in [115, 117, 74].

4.2 Construction of 1D Linear Stochastic PDEs

4.2.1 1D Linear Stochastic PDE Model

Without any a priori knowledge of the physics of a thin-film deposition process, we assume that there exists a one-dimensional linear stochastic PDE of the following general form that can adequately describe the evolution of the surface of a thin film during deposition:

$$\frac{\partial h}{\partial t} = c + c_0 h + c_1 \frac{\partial h}{\partial x} + c_2 \frac{\partial^2 h}{\partial x^2} + \cdots + c_w \frac{\partial^w h}{\partial x^w} + \xi(x, t), \tag{4.1}$$

where $x \in [0, \pi]$ is the spatial coordinate, t is the time, $h(x, t)$ is the height of the surface at position x and time t, and $\xi(x, t)$ is a Gaussian noise with zero mean and covariance

$$\langle \xi(x, t)\xi(x', t') \rangle = \varsigma^2 \delta(x - x')\delta(t - t'), \tag{4.2}$$

where $\delta(\cdot)$ is the Dirac function. Furthermore, the prederivative coefficients c and c_j in Eq. (4.1) and the parameter ς^2 in Eq. (4.2) depend on the process parameters (gas flow rates, substrate temperature, etc.) $p_i(t)$:

$$\begin{aligned} c &= C[p_1(t), p_2(t), \ldots, p_d(t)], \\ c_j &= C_j[p_1(t), p_2(t), \ldots, p_d(t)], \quad j = 0, \ldots, w, \\ \varsigma^2 &= C_\xi[p_1(t), p_2(t), \ldots, p_d(t)], \end{aligned} \tag{4.3}$$

where $C(\cdot)$, $C_j(\cdot)$, and $C_\xi(\cdot)$ are nonlinear functions to be determined.

The stochastic PDE in Eq. (4.1) is subjected to the following periodic boundary conditions:

$$\frac{\partial^j h}{\partial x^j}(0, t) = \frac{\partial^j h}{\partial x^j}(\pi, t), \quad j = 0, \ldots, w - 1, \tag{4.4}$$

and the initial condition

$$h(x, 0) = h_0(x). \tag{4.5}$$

Remark 4.1. In this work, we assume that a linear stochastic PDE model adequately describes the process dynamics. However, for cases in which the nonlinear dynamics are significant, nonlinear stochastic PDE models would be needed; this issue will be addressed in Section 4.4. Also, we note that we use a scalar function, $h(\cdot)$, to represent the height profile of the thin-film surface in the model. In general, $h(\cdot)$ can be a vector function and can be used to represent any appropriate microscopic description of the thin film (such as the defect locations, grain boundaries, etc.); in such a case, several stochastic PDEs should be considered simultaneously.

4.2.2 Eigenvalue Problem of the Linear Operator

To study the dynamics of Eq. (4.1), we initially consider the eigenvalue problem of the linear operator in Eq. (4.1), which takes the form

$$A\phi_n(x) = c_0 \phi_n(x) + c_1 \frac{d\phi_n(x)}{dx} + c_2 \frac{d^2 \phi_n(x)}{dx^2} + \ldots + c_w \frac{d^w \phi_n(x)}{dx^w} = \lambda_n \phi_n(x),$$

$$\frac{d^j \phi_n}{dx^j}(0) = \frac{d^j \phi_n}{dx^j}(\pi), \quad j = 0, \ldots, w - 1, \quad n = 1, \ldots, \infty, \tag{4.6}$$

where λ_n denotes an eigenvalue and ϕ_n denotes an eigenfunction. A direct computation of the solution of the above eigenvalue problem yields

$$\lambda_n = c_0 + I2nc_1 + (I2n)^2 c_2 + \cdots + (I2n)^w c_w,$$

$$\phi_n(x) = \sqrt{\frac{1}{\pi}} e^{I2nx}, \quad n = 0, \pm 1, \ldots, \pm \infty, \tag{4.7}$$

where λ_n denotes the nth eigenvalue, $\phi_n(x)$ denotes the nth eigenfunction, and $I = \sqrt{-1}$.

To present the method that we use for parameter identification of the stochastic PDE in Eq. (4.1), we first derive an infinite stochastic ODE representation of Eq. (4.1) using modal decomposition and parameterize the infinite stochastic ODE system using kMC simulation. We first expand the solution to Eq. (4.1) in an infinite series in terms of the eigenfunctions of the operator in Eq. (4.6) as follows (i.e., the Fourier expansion in the complex form):

$$h(x,t) = \sum_{n=-\infty}^{\infty} z_n(t)\phi_n(x), \tag{4.8}$$

where $z_n(t)$ are time-varying coefficients. Substituting the above expansion for the solution, $h(x,t)$, into Eq. (4.1) and taking the inner product, the following system of infinite stochastic ODEs is obtained:

$$\frac{dz_n}{dt} = \lambda_n z_n + c_{zn} + \xi_n(t), \quad n = 0, \pm 1, \ldots, \pm \infty, \tag{4.9}$$

with the initial conditions

$$z_n(0) = z_{n0}, \quad n = 0, \pm 1, \ldots, \pm \infty, \tag{4.10}$$

where $c_{zn} = c \int_0^\pi \phi_n^*(x)dx$ (note that $c_{z0} = c\sqrt{\pi}$ and $c_{zn} = 0 \; \forall \, n \neq 0$), $\xi_n(t) = \int_0^\pi \xi(x,t)\phi_n^*(x)dx$ and $z_{n0} = \int_0^\pi h_0(x)\phi_n^*(x)dx$, where $\phi_n^*(x)$ is the complex conjugate of $\phi_n(x)$, and the superscript star is used to denote complex conjugate in the remainder of this manuscript.

The covariances of $\xi_n(t)$ can be computed by using the following result:

Result 4.1. *If (1) $f(x)$ is a deterministic function, (2) $\eta(x)$ is a random variable with $\langle \eta(x) \rangle = 0$ and covariance $\langle \eta(x)\eta(x') \rangle = \sigma^2 \delta(x - x')$, and (3) $\epsilon = \int_a^b f(x)\eta(x)dx$, then ϵ is a random number with $\langle \epsilon \rangle = 0$ and covariance $\langle \epsilon^2 \rangle = \sigma^2 \int_a^b f(x)f^*(x)dx$* [1].

Using Result 4.1, we obtain $\langle \xi_n(t) \rangle = 0$ and $\langle \xi_n(t)\xi_n^*(t') \rangle = \varsigma^2 \delta(t-t')$. We note that $\xi_n(t)$ is a complex Gaussian random variable and the probability distribution function of the Gaussian distribution, $P(\xi_n, t)$, on the complex plane with zero mean and covariance $\varsigma^2 \delta(t - t')$ is defined as follows:

$$P(\xi_n, t) = \frac{1}{\sqrt{2\pi\varsigma\delta(t - t')}} e^{\xi_n \xi_n^* / 2\varsigma^2 \delta(t - t')}. \tag{4.11}$$

4.2.3 Analytical Solutions for Statistical Moments

To compute the parameters of the system of infinite stochastic ODEs in Eq. (4.9), we first derive the analytical expressions for the statistical moments of the stochastic ODE states, including the expected value and covariance. By comparing the analytical expression to the statistical moments obtained by multiple kMC simulations, the parameters of the stochastic ODE system (i.e., λ_n and ς) can be determined.

The explicit solution to Eq. (4.9) is obtained as follows to derive the expressions for the statistical moments of the stochastic ODE states:

$$z_n(t) = e^{\lambda_n t} z_{n0} + \frac{(e^{\lambda_n t} - 1)c_{zn}}{\lambda_n} + \int_0^t e^{\lambda_n(t - \mu)} \xi_n(\mu) d\mu. \tag{4.12}$$

Using Result 4.1, Eq. (4.12) can be further simplified as follows:

$$z_n(t) = e^{\lambda_n t} z_{n0} + \frac{(e^{\lambda_n t} - 1)c_{zn}}{\lambda_n} + \theta_n(t), \tag{4.13}$$

where $\theta_n(t)$ is a complex random variable of normal distribution with zero mean and covariance $\langle \theta_n(t)\theta_n^*(t) \rangle = \varsigma^2 (e^{(\lambda_n + \lambda_n^*)t} - 1)/(\lambda_n + \lambda_n^*)$. Therefore, the expected value (the first stochastic moment) and the covariance (the second stochastic moment) of state z_n can be expressed as follows:

$$\langle z_n(t) \rangle = e^{\lambda_n t} z_{n0} + \frac{(e^{\lambda_n t} - 1)c_{zn}}{\lambda_n},$$

$$\langle z_n(t)z_n^*(t) \rangle = \varsigma^2 \frac{e^{(\lambda_n + \lambda_n^*)t} - 1}{\lambda_n + \lambda_n^*} + \langle z_n(t) \rangle \langle z_n(t) \rangle^*, \tag{4.14}$$

$$n = 0, \pm 1, \ldots, \pm \infty.$$

Equation (4.14) holds for any initial condition z_{n0}. Since we are able to choose any initial thin-film surface for simulation, we choose $z_{n0} = 0$ [i.e., the initial surface is flat, $h(x, 0) = 0$] to simplify our calculations. In this case, Eq. (4.14) can be further simplified as follows (note that $c_{zn} = 0, \forall n \neq 0$):

$$\langle z_n(t) \rangle = 0,$$

$$\langle z_n(t)z_n^*(t) \rangle = \varsigma^2 \frac{e^{(\lambda_n + \lambda_n^*)t} - 1}{\lambda_n + \lambda_n^*} = \varsigma^2 \frac{e^{2\mathrm{Re}(\lambda_n)t} - 1}{2\mathrm{Re}(\lambda_n)}, \tag{4.15}$$

$$n = \pm 1, \ldots, \pm \infty,$$

where $\mathrm{Re}(\lambda_n)$ denotes the real part of λ_n, and for $z_0(t)$, it follows from Eq. (4.14) with $\lambda_0 = 0$ that

$$\langle z_0(t) \rangle = \lim_{\lambda_0 \to 0} \frac{(e^{\lambda_0 t} - 1)c_{z0}}{\lambda_0} = tc_{z0} = t\sqrt{\pi}c, \tag{4.16}$$

$$\langle z_0^2(t) \rangle = \varsigma^2 t + t^2 \pi c^2.$$

It can be seen in Eq. (4.15) that the statistical moments of each stochastic ODE state depend only on the real part of the corresponding eigenvalue. Therefore, to determine the imaginary part of the eigenvalue, we need to construct an additional equation. We note that λ_n would be a complex number if the linear operator A is not self-adjoint, i.e., when odd partial derivatives are present in the stochastic PDE [see Eq. (4.7)].

Therefore, we rewrite Eq. (4.12) by separating the real part and the imaginary part of $z_n(t)$ as follows with initial condition $z_{n0} = 0$:

$$z_n(t) = \frac{1}{2} \int_0^t [e^{\lambda_n(t-\mu)} + e^{\lambda_n^*(t-\mu)}] \xi_n(\mu) d\mu$$
$$+ \frac{1}{2} \int_0^t [e^{\lambda_n(t-\mu)} - e^{\lambda_n^*(t-\mu)}] \xi_n(\mu) d\mu, \tag{4.17}$$
$$n = \pm 1, \ldots, \pm\infty.$$

Accordingly, the real part of $z_n(t)$ can be expressed as follows:

$$\text{Re}[z_n(t)] = \frac{1}{2} \int_0^t [e^{\lambda_n(t-\mu)} + e^{\lambda_n^*(t-\mu)}] \xi_n(\mu) d\mu, \quad n = \pm 1, \ldots, \pm\infty, \tag{4.18}$$

where $\text{Re}[z_n(t)]$ denotes the real part of $z_n(t)$. Using Result 4.1, we have

$$\langle \text{Re}[z_n(t)] \rangle = 0,$$
$$\langle \text{Re}[z_n(t)]^2 \rangle = \varsigma^2 \left[\frac{\lambda_n^* e^{2\lambda_n t} + \lambda_n e^{2\lambda_n^* t} - (\lambda_n + \lambda_n^*)}{8\lambda_n \lambda_n^*} + \frac{e^{(\lambda_n + \lambda_n^*)t} - 1}{2(\lambda_n + \lambda_n^*)} \right]$$
$$= \varsigma^2 \left\{ \frac{\text{Re}(\lambda_n) e^{2\text{Re}(\lambda_n)} \cos(2\text{Im}(\lambda_n)t)}{4[\text{Re}(\lambda_n)^2 + \text{Im}(\lambda_n)^2]} \right.$$
$$+ \frac{\text{Im}(\lambda_n) e^{2\text{Re}(\lambda_n)} \sin(2\text{Im}(\lambda_n)t)}{4[\text{Re}(\lambda_n)^2 + \text{Im}(\lambda_n)^2]}$$
$$\left. - \frac{\text{Re}(\lambda_n)}{4[\text{Re}(\lambda_n)^2 + \text{Im}(\lambda_n)^2]} + \frac{e^{2\text{Re}(\lambda_n)t} - 1}{4\text{Re}(\lambda_n)} \right\},$$
$$n = \pm 1, \ldots, \pm\infty,$$

(4.19)

where $\text{Im}(\lambda_n)$ denotes the imaginary part of λ_n. Thus, we can use Eq. (4.15) to first determine the real part of the eigenvalue, and then use Eq. (4.19) to determine its imaginary part. We note that it is not recommended to determine both parts of the eigenvalue using only Eq. (4.19), since, in that case, the nonlinear least-squares problem involved in the eigenvalue determination would be much more difficult to solve.

4.2.4 Model Construction Methodology

Equations (4.15), (4.16), and (4.19) show the analytical relationship that relates the linear operator and the Gaussian noise in Eq. (4.1) to the statistical moments of the states of Eq. (4.9) which can be obtained, through multiple experimental measurements or kinetic Monte Carlo simulations and, therefore, reveal a viable path to systematically construct a linear stochastic PDE of the form of Eq. (4.1) that describes the dynamics of microscopic processes directly from experimental or simulation data. To this end, we propose a systematic procedure to construct linear stochastic PDEs. This procedure will be demonstrated using a thin-film deposition process described in Section 4.2.5 and can be readily extended to other stochastic processes.

The proposed procedure includes the following steps: First, we design a set of simulation experiments that cover the complete range of process operation; second, we run multiple simulations for each simulation condition to obtain the trajectories of the first and second statistical moments of the states (i.e., Fourier coefficients) computed from the surface snapshots; third, we compute the eigenvalues of the linear operator and covariance of the Gaussian noise based on the trajectories of the statistical moments of the states for each simulation run, and determine the model parameters of the stochastic PDE (i.e., the prederivative coefficients and the order of the stochastic PDE); finally, we investigate the dependence of the model parameters of the stochastic PDE on the process parameters and determine the least-squares optimal form of the stochastic PDE model with model parameters expressed as functions of the process parameters.

4.2.5 Application to a 1D Thin-Film Growth Process

To illustrate the application of the above model construction methodology, we consider a thin-film growth process of deposition from the vapor phase in which the formation of the thin film is governed by two microscopic processes that occur on the surface as shown in Fig. 4.1, i.e., the adsorption of vapor-phase molecules on the surface and the migration of surface molecules. The processes of molecule adsorption and migration are very common in thin-film growth processes.

More specifically, we consider a single-species growth on a one-dimensional lattice (see Section 4.3 for extension of the results to thin-film growth occurring in a two-dimensional lattice). The adsorption rate, which depends on the vapor-phase concentration, is considered uniform over the spatial domain and constant (i.e., fixed growth rate) during each deposition. However, it could vary for different deposition runs. All surface sites are available for adsorption for all times and the adsorption rate for each surface site is given by

$$w_a = W, \tag{4.20}$$

where W is the growth rate in ML/s (monolayers per second).

The migration rate of each surface molecule depends on its local environment. Under the consideration of only first nearest-neighbor interactions, the migration rate of surface molecules from a surface site with n first nearest neighbors is given by

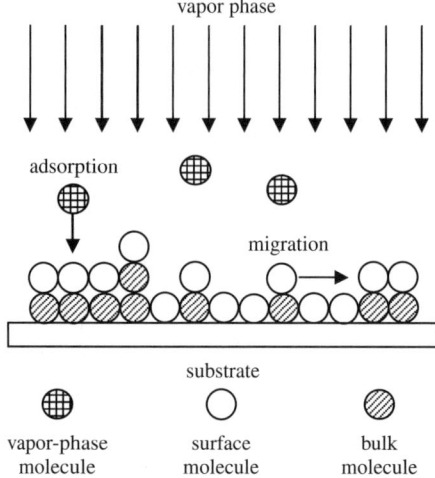

Fig. 4.1. The thin film growth process.

$$w_m(n) = k_{m0}e^{-\frac{E_s+nE_n}{k_BT}}, \tag{4.21}$$

where E_s is the energy barrier associated with migration due to surface effects, E_n is the energy barrier associated with migration due to nearest-neighbor interactions, k_{m0} is the frequency constant associated with migration, k_B is Boltzmann's constant, and T is the substrate temperature. The values of migration energy barriers and frequency constant used in this study are taken from the literature [133] for a molecular-beam epitaxy GaAs process and are as follows: $E_s = 1.58\,\mathrm{eV}$, $E_n = 0.28\,\mathrm{eV}$, and $k_{m0} = 2k_BT/h$, where h is Planck's constant. A kinetic Monte Carlo simulation code is used to simulate the deposition process and obtain surface snapshots.

Eigenvalues and Covariance

Because only two process parameters are considered in the deposition process under consideration, the growth rate W and the substrate temperature T, the design of the simulation runs is straightforward. Specifically, different W values and T values are evenly selected from the range of process operations of interest and simulation runs are executed with every selected W value for each selected T value. Therefore, we start our demonstration of the model construction methodology with the computation of the eigenvalues and of the covariance. Also, we note that the trajectories of the statistical moments for each simulation condition are computed based on 100 simulation runs taking place with the same process parameters.

In the previous section, we showed that for a deposition process with a flat initial surface, the covariance of each state $\langle z_n(t)z_n^*(t)\rangle$ can be predicted by Eq. (4.15). Therefore, we can fit ς^2 and $\mathrm{Re}(\lambda_n)$ in Eq. (4.15) for the profile of $\langle z_n(t)z_n^*(t)\rangle$.

In order to obtain the profile of $\langle z_n(t)z_n^*(t)\rangle$, we need to generate snapshots of the thin-film surface during each deposition simulation and compute the values of

$z_n(t)$. Since the lattice consists of discrete sites, we let $h(kL,t)$ be the height profile of the surface at time t with lattice constant L (k denotes the coordinate of a specific surface site), and compute $z_n(t)$ as follows:

$$z_n(t) = \int_0^\pi h(x,t)\phi_n^*(x)dx = \sum_{k=0}^{k_{max}} h(kL,t)\int_{kL}^{(k+1)L} \phi_n^*(x)dx, \qquad (4.22)$$

where $k_{max}L = \pi$ (i.e., the lattice is mapped to the domain $[0,\pi]$). Substituting Eq. (4.7) into Eq. (4.22), we can derive the following expression for $z_n(t)$:

$$z_n(t) = \sum_{k=0}^{k_{max}} \frac{h(kL,t)e^{-I2kLn}}{I2\sqrt{\pi n}}(1 - e^{-I2Ln}), \quad n = \pm 1, \ldots, \pm\infty, \qquad (4.23)$$

and for $z_0(t)$, we have

$$z_0(t) = \sum_{k=0}^{k_{max}} h(kL,t)\frac{L}{\sqrt{\pi}} = \sqrt{\pi}t\frac{\displaystyle\sum_{k=0}^{k_{max}} h(kL,t)}{k_{max}t} = t\sqrt{\pi}W. \qquad (4.24)$$

To capture the dynamics of both the fast states and slow states simultaneously in the same simulation run with few surface snapshots, the snapshots are generated in a variable time-step fashion in which the intervals between two snapshots are increased with time. This procedure is motivated by the fact that the dynamics of the fast states can be detected only at the beginning of each simulation run, and, therefore, the evolving surface should be sampled more frequently in the beginning than during the remainder to cope with the small time scale of evolution of these fast states.

Figure 4.2 shows the eigenvalues computed from thin-film depositions occurring under the same operating conditions but simulated with different lattice sizes (we note that the computed eigenvalues are considered real since the imaginary part of the eigenvalues turned out to be very small, which implies the absence of odd-order terms in the stochastic PDE model of this process). It can be seen that the computed spectrums are very close to each other when n is rescaled with the corresponding lattice size. This is expected, since $\phi_n(x)$ is a basis of the domain of operator A and is a complex function of the frequency n. Accordingly, n/k_{max} is the length scale of the surface fluctuation described by $\phi_n(x)$ when a lattice of size k_{max} is mapped to the domain of $[0,\pi]$ (we note that, for the same reason, the covariance values should be scaled with the inverse of the lattice size, $1/k_{max}$, in order to carry out a meaningful comparison).

It can also be seen in Fig. 4.2 that the eigenspectrums are very close to the parabolic reference curve, which implies that a second-order stochastic PDE system of the following form would be able to describe the evolution of the surface height of this deposition process:

$$\frac{\partial h}{\partial t} = c + c_2\frac{\partial^2 h}{\partial x^2} + \xi(x,t), \qquad (4.25)$$

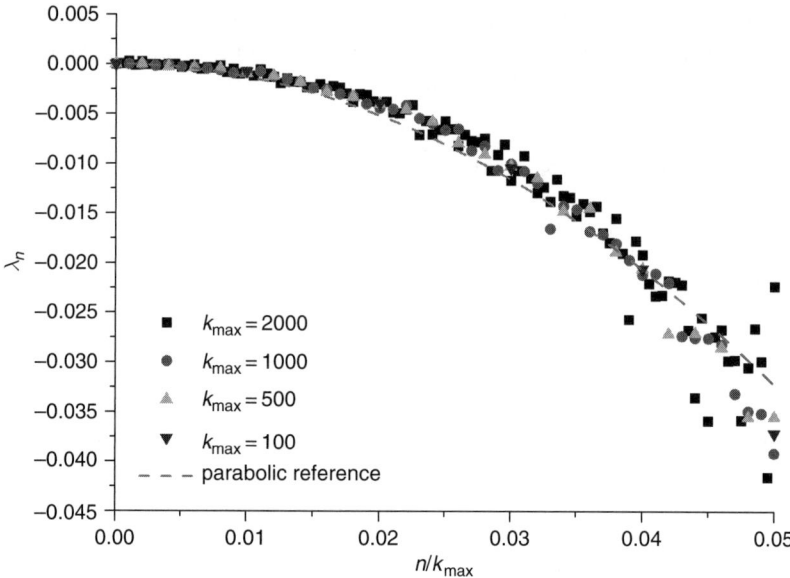

Fig. 4.2. Eigenvalue spectrums of the system of infinite stochastic ODEs computed from the kMC simulation of the deposition process with different lattice sizes: $k_{max} = 100$, $k_{max} = 500$, $k_{max} = 1000$, and $k_{max} = 2000$.

where c, c_2, and the covariance of the Gaussian noise ξ, ς, all depend on the microscopic processes and operating conditions.

Dependence on the Process Parameters

We now proceed with the derivation of the parameters of the stochastic PDE of Eq. (4.25). From Eqs. (4.16) and (4.24), we can see that $c = W$ for all cases. However, c_2 and ς^2 identified for different deposition settings can be very different, and so we need to investigate their dependence on the deposition parameters to obtain their analytical expressions. c_2 and ς^2 are evaluated for assorted deposition conditions, and a lattice size of 1000 (i.e., $k_{max} = 1000$) is used for all the simulation runs in our study.

Figure 4.3 shows the eigenspectrums and the covariance spectrums computed from depositions with the same growth rate ($W = 0.5\,\mathrm{ML/s}$) for different substrate temperatures. It can be seen that the magnitude of the eigenvalues decreases faster with increasing n at higher substrate temperatures. This implies that a higher substrate temperature corresponds to a larger c_2 in the stochastic PDE model, and vice versa. Although it follows from Eq. (4.11) that the covariance of the stochastic noise should be the same for all states, it is not exactly the case for high-order states in the high-substrate-temperature regime (e.g., $T = 680\,\mathrm{K}$). However, because these high-order states correspond to the surface fluctuations of small length scales, and at the same time, such small length-scale surface fluctuations are almost negligible in the

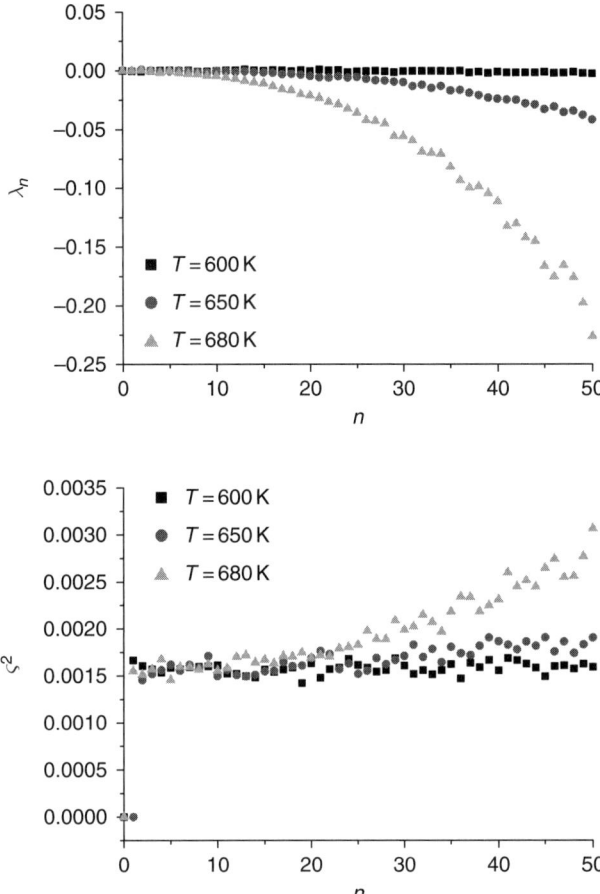

Fig. 4.3. Eigenspectrums (top) and covariance spectrums (bottom) computed from simulated deposition processes with a growth rate $W = 0.5\,\mathrm{ML/s}$ for different substrate temperatures: $T = 600\,\mathrm{K}$, $T = 650\,\mathrm{K}$, and $T = 680\,\mathrm{K}$.

high-substrate-temperature regime due to the significant surface diffusion, the contribution from these high-order states at high substrate temperatures becomes very small. Therefore, given that such discrepancy would not significantly affect the accuracy of the model, we compute ς^2 only based on the low-order states. From the covariance of the low-order states shown in the right plot of Fig. 4.3, we may also consider ς^2 to be independent of substrate temperature.

We also investigate the eigenspectrums and covariance spectrums computed from depositions occurring under the same substrate temperature but different thin-film growth rates. It can be seen that a higher growth rate corresponds to a smaller c_2 but a larger covariance value [118].

To derive explicit expressions for c_2 and ς^2 as functions of T and W, we evaluate these values for different T and W [118]. We find that $\ln c_2$ has a quasi-linear relationship with both T and W. Thus, the following expression can be obtained for c_2 as a function of T and W through least-squares fitting:

$$
\begin{aligned}
c_2(T, W) &= e^{-45.8176 + 0.0511T - 0.1620W} \\
&= \frac{e^{-32.002 + 0.0511T - 0.1620W}}{k_{\max}^2}.
\end{aligned}
\tag{4.26}
$$

Furthermore, we find that ς^2 depends almost linearly on both T and W. Thus, the following expression can be obtained for ς^2 as a function of T and W through least-squares fitting as well:

$$
\varsigma^2(T, W) = 5.137 \times 10^{-8}T + 3.2003 \times 10^{-3}W \approx \frac{\pi W}{k_{\max}}.
\tag{4.27}
$$

Therefore, the linear stochastic PDE model computed for the deposition process is as follows:

$$
\frac{\partial h}{\partial t} = W + \left(\frac{e^{-32.002 + 0.0511T - 0.1620W}}{k_{\max}^2} \right) \frac{\partial^2 h}{\partial x^2} + \xi(x, t),
$$

$$
\frac{\partial h}{\partial x}(0, t) = \frac{\partial h}{\partial x}(\pi, t), \quad h(0, t) = h(\pi, t), \quad h(x, 0) = h_0(x),
\tag{4.28}
$$

where

$$
\langle \xi(x, t)\xi(x', t') \rangle = \frac{5.137 \times 10^{-5}T + 3.2003W}{k_{\max}} \delta(x - x')\delta(t - t').
$$

Validation of the 1D Stochastic PDE Model

We now proceed with the validation of the stochastic PDE model of the thin-film deposition process [Eq. (4.28)]. Validation experiments are conducted for a number of deposition conditions that have not been used for the model construction. We generate surface profiles using both the stochastic PDE model and the kinetic Monte Carlo code. Figure 4.4 shows the surface profile at the end of a deposition with substrate temperature $T = 550\,\text{K}$, thin-film growth rate $W = 0.1\,\text{ML/s}$, deposition duration of $1000\,\text{s}$, and lattice size $k_{\max} = 2000$. Figure 4.5 shows the surface profile at the end of a deposition with substrate temperature $T = 700\,\text{K}$, thin-film growth rate $W = 2.5\,\text{ML/s}$, deposition duration of $400\,\text{s}$ and lattice size $k_{\max} = 2000$; we can see that at both low and high substrate temperatures, and for different growth rates, the surface height profile, $h(x)$, at $t = 1000\,\text{s}$ computed by the linear stochastic PDE model constructed for the deposition process is very consistent with that obtained from the kinetic Monte Carlo simulation.

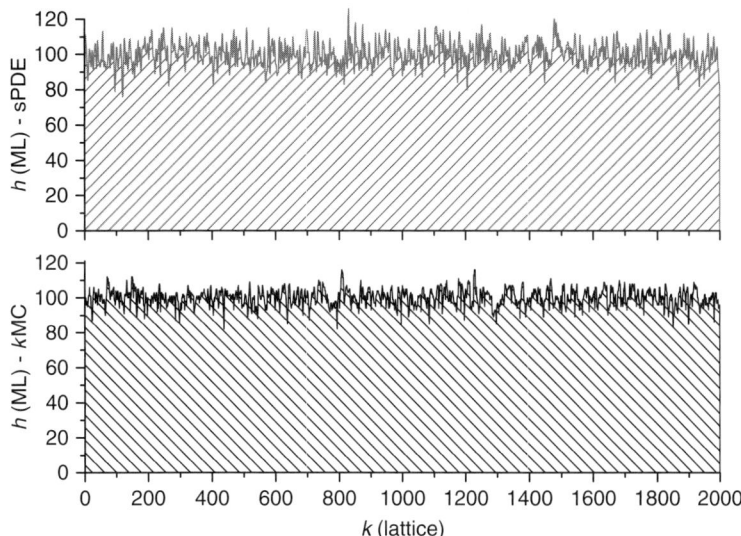

Fig. 4.4. Final thin-film surface profiles generated by kMC simulation and stochastic PDE model for a 1000-s deposition with substrate temperature $T = 550$ K, thin-film growth rate $W = 0.1$ ML/s and lattice size $k_{\max} = 2000$.

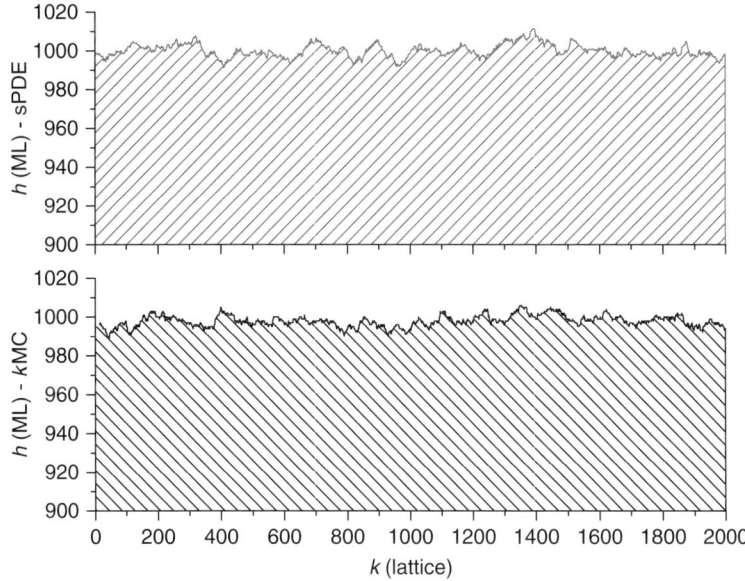

Fig. 4.5. Final thin-film surface profiles generated by kMC simulation and stochastic PDE model for a 400-s deposition with substrate temperature $T = 700$ K, thin-film growth rate $W = 2.5$ ML/s, and lattice size $k_{\max} = 2000$.

We also generate expected surface roughness profiles using both the stochastic PDE model and the kinetic Monte Carlo simulation (average of 100 runs) for the deposition process. For simplicity, the surface roughness is evaluated in a root-mean-square fashion as follows:

$$r(t) = \sqrt{\frac{1}{\pi} \int_0^{\pi} [h(x,t) - \bar{h}(t)]^2 dx,} \tag{4.29}$$

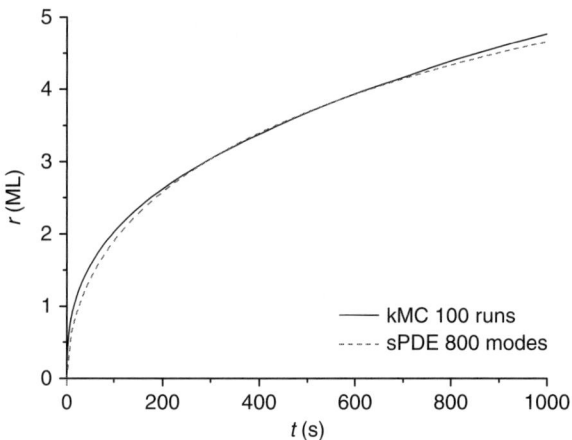

Fig. 4.6. Expected surface roughness profiles generated by kMC simulation and stochastic PDE model for a 1000-s deposition with substrate temperature $T = 550$ K, thin-film growth rate $W = 0.1$ ML/s, and lattice size $k_{max} = 2000$.

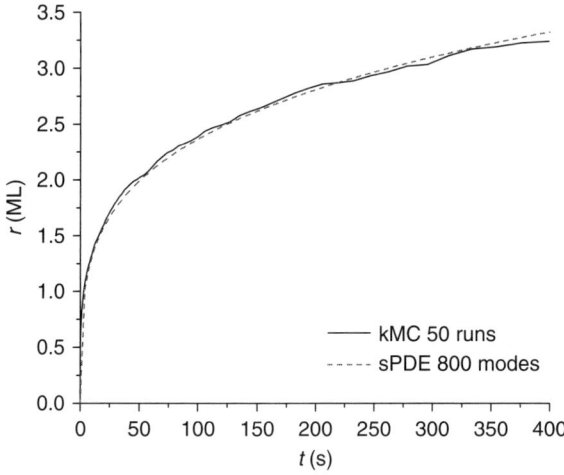

Fig. 4.7. Expected surface roughness profiles generated by kMC simulation and stochastic PDE model for a 400-s deposition with substrate temperature $T = 700$ K, thin-film growth rate $W = 2.5$ ML/s, and lattice size $k_{max} = 2000$.

where $\bar{h}(t) = (1/\pi) \int_0^\pi h(x,t)dx$ is the average surface height. We note that for a more detailed description of the surface morphology, the height–height correlation function may be used to evaluate the surface roughness [144].

Figure 4.6 shows the expected roughness profile of a deposition with substrate temperature $T = 550\,$K and thin-film growth rate W $= 0.1\,$ML/s; Fig.4.7 shows the roughness profile of a deposition with substrate temperature $T = 700\,$K and thin-film growth rate $W = 2.5\,$ML/s; we can see that the linear stochastic PDE model constructed for the deposition process is also very consistent with the kinetic Monte Carlo simulation in terms of surface roughness prediction, at both low and high substrate temperatures, for different growth rates.

4.3 Construction of 2D Linear Stochastic PDEs

In Section 4.2, a systematic method to construct stochastic PDE models for thin-film growth using first-principles-based microscopic simulations has been presented, and the method has been applied to a representative one-dimensional thin-film growth process. In this section, we focus on the construction of two-dimensional linear stochastic PDE models for two-dimensional thin-film growth processes.

4.3.1 2D Linear Stochastic PDE Model

We now proceed with constructing a closed-form stochastic PDE model using the approach we presented in Section 4.2. Without any a priori knowledge of the deposition process, we assume that there exists a 2D linear stochastic PDE of the following general form that can adequately describe the evolution of the surface of the thin film during the deposition:

$$\frac{\partial h}{\partial t} = c + c_1 \nabla h + c_2 \nabla^2 h + \cdots + c_w \nabla^w h + \xi(x,y,t), \tag{4.30}$$

where $x \in [0,\pi]$, $y \in [0,\pi]$ is the spatial coordinate, t is the time, $h(x,y,t)$ is the height (in the unit of $monolayers$) of the surface at position x,y and time t, ∇^k is the following operator: $\nabla^k = \frac{\partial^k}{\partial x^k} + \frac{\partial^k}{\partial y^k}$, and $\xi(x,y,t)$ is a Gaussian noise with zero mean and covariance:

$$\langle \xi(x,y,t)\xi(x',y',t')\rangle = \varsigma^2 \delta(x-x')\delta(y-y')\delta(t-t'), \tag{4.31}$$

where $\delta(\cdot)$ is the Dirac function. Furthermore, the prederivative coefficients c and c_j in Eq. (4.30) and the parameter ς^2 in Eq. (4.31) depend on the process parameters, the substrate temperature T, and the adsorption rate W (directly determined by the vapor-phase concentration):

$$\begin{aligned} c &= C[T(t), W(t)], \\ \varsigma^2 &= C_\xi[T(t), W(t)], \\ c_j &= C_j[T(t), W(t)], \quad j = 0, \ldots, w, \end{aligned} \tag{4.32}$$

where $C(\cdot)$ and $C_j(\cdot)$ are nonlinear functions to be determined.

The stochastic PDE in Eq. (4.30) is subjected to the following periodic boundary conditions:

$$\nabla^j h(0, y, t) = \nabla^j h(\pi, y, t),$$
$$\nabla^j h(x, 0, t) = \nabla^j h(x, \pi, t), \quad j = 0, \ldots, w - 1, \tag{4.33}$$

with the initial condition

$$h(x, y, 0) = h_0(x, y). \tag{4.34}$$

4.3.2 Eigenvalue Problem of the Linear Operator

To study the dynamics of Eq. (4.30), we initially consider the eigenvalue problem of the linear operator of Eq. (4.30), which takes the form

$$A\phi_{m,n}(x, y) \; = c_1 \nabla \phi_{m,n}(x, y) + c_2 \nabla^2 \phi_{m,n}(x, y) + \cdots + c_w \nabla^w \phi_{m,n}(x, y)$$

$$= \lambda_{m,n}\phi_{m,n}(x, y),$$

$$\nabla^j \phi_{m,n}(0, y) = \nabla^j \phi_{m,n}(\pi, y),$$

$$\nabla^j \phi_{m,n}(x, 0) = \nabla^j \phi_{m,n}(x, \pi), \quad j = 0, \ldots, w - 1, \quad m, n = 0, \pm 1, \ldots, \pm\infty, \tag{4.35}$$

where $\lambda_{m,n}$ denotes the eigenvalue and $\phi_{m,n}$ denotes the eigenfunction. A direct computation of the solution to the above eigenvalue problem yields

$$\lambda_{m,n} = (I2m + I2n)c_1 + [(I2m)^2 + (I2n)^2]c_2$$
$$+ \cdots + [(I2m)^w + (I2n)^w]c_w, \tag{4.36}$$

$$\phi_{m,n}(x, y) = \frac{1}{\pi}(e^{I2mx + I2ny}), \quad m, n = 0, \pm 1, \ldots, \pm\infty,$$

where $\lambda_{m,n}$ denotes the (m, n)th eigenvalue, $\phi_{m,n}(x, y)$ denotes the (m, n)th eigenfunction, and $I = \sqrt{-1}$.

To present the method that we use for the identification of the stochastic PDE in Eq. (4.30), we first derive an infinite stochastic ODE representation of Eq. (4.30) using modal decomposition and parameterize the infinite stochastic ODE system using kMC simulation. We first expand the solution to Eq. (4.30) in an infinite series in terms of the eigenfunctions of the operator in Eq. (4.35) as follows (i.e., the Fourier expansion in the complex form):

$$h(x, y, t) = \sum_{m,n=-\infty}^{\infty} z_{m,n}(t)\phi_{m,n}(x, y), \tag{4.37}$$

where $z_{m,n}(t)$ are time-varying coefficients. Substituting the above expansion for the solution, $h(x, y, t)$, into Eq. (4.30) and taking the inner product, the following system of infinite stochastic ODEs is obtained:

$$\frac{dz_{m,n}}{dt} = \lambda_{m,n}z_{m,n} + c_{m,n}^z + \xi_{m,n}(t), \quad m, n = 0, \pm 1, \ldots, \pm\infty, \tag{4.38}$$

with the initial conditions

$$z_{m,n}(0) = z_{m,n,0}, \quad m, n = 0, \pm 1, \ldots, \pm\infty, \tag{4.39}$$

where $c_{m,n}^z = c\int_0^\pi \int_0^\pi \phi_{m,n}^*(x, y)dxdy$, $\xi_{m,n}(t) = \int_0^\pi \int_0^\pi \xi(x, y, t)\phi_{m,n}^*(x, y)\,dxdy$, and $z_{m,n,0} = \int_0^\pi \int_0^\pi h_0(x, y)\phi_{m,n}^*(x, y)dxdy$. We note that $c_{0,0}^z = \pi c$ and $c_{m,n}^z = 0$ when $m^2 + n^2 \neq 0$. $\phi_{m,n}^*$ is the complex conjugate of $\phi_{m,n}$, and the superscript star is used to denote complex conjugate in the remainder of this section.

By using Result 4.1, we obtain $\langle \xi_{m,n}(t) \rangle = 0$ and $\langle \xi_{m,n}(t)\xi_{m,n}^*(t') \rangle = \varsigma^2 \delta(t - t')$. We note that $\xi_{m,n}(t)$ is a complex Gaussian random variable.

4.3.3 Analytical Solutions for Statistical Moments

To compute the parameters of the system of infinite stochastic ODEs in Eq. (4.38), we first derive the analytical expressions for the statistical moments of the stochastic ODE states, such as the expected values and covariances. By comparing the analytical expression to the statistical moments obtained by multiple kMC simulations, the parameters of the stochastic ODE system (i.e., $\lambda_{m,n}$ and ς) can be determined.

The analytical solution to Eq. (4.38) is obtained as follows to derive the expressions for the statistical moments of the stochastic ODE states:

$$z_{m,n}(t) = e^{\lambda_{m,n}t}z_{m,n,0} + \frac{(e^{\lambda_{m,n}t} - 1)c_{m,n}^z}{\lambda_{m,n}} + \int_0^t e^{\lambda_{m,n}(t - \mu)}\xi_{m,n}(\mu)d\mu, \tag{4.40}$$

Using Result 4.1, Eq. (4.40) can be further simplified as follows:

$$z_{m,n}(t) = e^{\lambda_{m,n}t}z_{m,n,0} + \frac{(e^{\lambda_{m,n}t} - 1)c_{m,n}^z}{\lambda_{m,n}} + \theta_{m,n}(t), \tag{4.41}$$

where $\theta_{m,n}(t)$ is a complex random variable of normal distribution with zero mean and covariance $\langle \theta_{m,n}(t)\theta_{m,n}^*(t) \rangle = \varsigma^2(e^{(\lambda_{m,n} + \lambda_{m,n}^*)t} - 1)/(\lambda_{m,n} + \lambda_{m,n}^*)$. Therefore, the first stochastic moment (the expected value) and the second stochastic moment (the covariance) of state $z_{m,n}$ can be expressed as follows:

$$\langle z_{m,n}(t) \rangle = e^{\lambda_{m,n}t}z_{m,n,0} + \frac{(e^{\lambda_{m,n}t} - 1)c_{m,n}^z}{\lambda_{m,n}},$$

$$\langle z_{m,n}(t)z_{m,n}^*(t) \rangle = \varsigma^2 \frac{e^{(\lambda_{m,n} + \lambda_{m,n}^*)t} - 1}{\lambda_{m,n} + \lambda_{m,n}^*} + \langle z_{m,n}(t) \rangle \langle z_{m,n}(t) \rangle^*. \tag{4.42}$$

Remark 4.2. We note that Eqs. (4.40), (4.41), and (4.42) hold for all stochastic ODE system states. Particularly, when $m = n = 0$ (i.e., for state $z_{0,0}$), $\lambda_{m,n} = 0$, these terms in the equations with $\lambda_{m,n}$ and $\lambda_{m,n} + \lambda_{m,n}^*$ as denominators should be calculated as follows:

$$\lim_{\lambda_{m,n} \to 0} \frac{(e^{\lambda_{m,n}t} - 1)c_{m,n}^z}{\lambda_{m,n}} = tc_{m,n}^z,$$

$$\lim_{\lambda_{m,n} \to 0} \varsigma^2 \frac{e^{(\lambda_{m,n} + \lambda_{m,n}^*)t} - 1}{\lambda_{m,n} + \lambda_{m,n}^*} = \varsigma^2 t.$$

(4.43)

Equation (4.42) holds for any initial condition $z_{m,n,0}$. Since we are able to choose any initial thin-film surface profile for simulation, we choose $z_{m,n,0} = 0$ [i.e., the initial surface is flat, $h(x, y, 0) = 0$] to simplify our calculations. In this case, Eq. (4.42) can be further simplified as follows (note that $c_{m,n}^z = 0, \forall m^2 + n^2 \neq 0$):

$$\langle z_{m,n}(t) \rangle = 0,$$

$$\langle z_{m,n}(t) z_{m,n}^*(t) \rangle = \varsigma^2 \frac{e^{(\lambda_{m,n} + \lambda_{m,n}^*)t} - 1}{\lambda_{m,n} + \lambda_{m,n}^*} = \varsigma^2 \frac{e^{2\mathrm{Re}(\lambda_{m,n})t} - 1}{2\mathrm{Re}(\lambda_{m,n})}, \quad (4.44)$$

$$m, n = 0, \pm 1, \ldots, \pm \infty, \quad m^2 + n^2 \neq 0,$$

where $\mathrm{Re}(\lambda_{m,n})$ denotes the real part of $\lambda_{m,n}$. For $z_{0,0}(t)$, it follows from Eq. (4.42) with $\lambda_{0,0} = 0$ that

$$\langle z_{0,0}(t) \rangle = t\pi c,$$

$$\langle z_{0,0}^2(t) \rangle = \varsigma^2 t + t^2 \pi^2 c^2.$$

(4.45)

It can be seen in Eq. (4.44) that the statistical moments of each stochastic ODE state depend only on the real part of the corresponding eigenvalue. Therefore, to determine the imaginary part of the eigenvalue, we need to construct an extra equation. We note that $\lambda_{m,n}$ will be a complex number if the linear operator A is not self-adjoint, for example, when odd partial derivatives are present in the stochastic PDE [see Eq. (4.36)].

Therefore, we rewrite Eq. (4.40) by separating the real part and the imaginary part of $z_{m,n}(t)$ as follows with initial condition $z_{m,n,0} = 0$:

$$z_{m,n}(t) = \frac{1}{2} \int_0^t [e^{\lambda_{m,n}(t - \mu)} + e^{\lambda_{m,n}^*(t - \mu)}] \xi_{m,n}(\mu) d\mu$$

$$+ \frac{1}{2} \int_0^t [e^{\lambda_{m,n}(t - \mu)} - e^{\lambda_{m,n}^*(t - \mu)}] \xi_{m,n}(\mu) d\mu, \quad (4.46)$$

$$m, n = 0, \pm 1, \ldots, \pm \infty, \quad m^2 + n^2 \neq 0.$$

Accordingly, the real part of $z_{m,n}(t)$ can be expressed as follows:

$$\mathrm{Re}[z_{m,n}(t)] = \frac{1}{2} \int_0^t [e^{\lambda_{m,n}(t - \mu)} + e^{\lambda_{m,n}^*(t - \mu)}] \xi_{m,n}(\mu) d\mu, \quad (4.47)$$

$$m, n = 0, \pm 1, \ldots, \pm \infty, \quad m^2 + n^2 \neq 0,$$

where $\mathrm{Re}[z_{m,n}(t)]$ denotes the real part of $z_{m,n}(t)$. By using Result 4.1, we have

$$\langle \mathrm{Re}[z_{m,n}(t)]\rangle = 0,$$

$$\langle \mathrm{Re}[z_{m,n}(t)]^2\rangle = \varsigma^2 \left[\frac{\lambda_{m,n}^* e^{2\lambda_{m,n}t} + \lambda_{m,n} e^{2\lambda_{m,n}^* t} - (\lambda_{m,n} + \lambda_{m,n}^*)}{8\lambda_{m,n}\lambda_{m,n}^*} \right.$$

$$\left. + \frac{e^{(\lambda_{m,n} + \lambda_{m,n}^*)t} - 1}{2(\lambda_{m,n} + \lambda_{m,n}^*)} \right]$$

$$= \varsigma^2 \left\{ \frac{\mathrm{Re}(\lambda_{m,n})e^{2\mathrm{Re}(\lambda_{m,n})}\cos(2\mathrm{Im}(\lambda_{m,n})t)}{4[\mathrm{Re}(\lambda_{m,n})^2 + \mathrm{Im}(\lambda_{m,n})^2]} \right. \tag{4.48}$$

$$+ \frac{\mathrm{Im}(\lambda_{m,n})e^{2\mathrm{Re}(\lambda_{m,n})}\sin(2\mathrm{Im}(\lambda_{m,n})t)}{4[\mathrm{Re}(\lambda_{m,n})^2 + \mathrm{Im}(\lambda_{m,n})^2]}$$

$$\left. - \frac{\mathrm{Re}(\lambda_{m,n})}{4[\mathrm{Re}(\lambda_{m,n})^2 + \mathrm{Im}(\lambda_{m,n})^2]} + \frac{e^{2\mathrm{Re}(\lambda_{m,n})t} - 1}{4\mathrm{Re}(\lambda_{m,n})} \right\},$$

$$m, n = 0, \pm 1, \ldots, \pm\infty, \quad m^2 + n^2 \neq 0,$$

where $\mathrm{Im}(\lambda_{m,n})$ denotes the imaginary part of $\lambda_{m,n}$. Thus, we can use Eq. (4.44) to first determine the real part of the eigenvalue, and then use Eq. (4.48) to determine its imaginary part. We note that we can determine both parts of the eigenvalue using only Eq. (4.48). However, in that case, the nonlinear least-squares problem involved in the eigenvalue determination would be much more difficult to solve.

4.3.4 Model Construction Methodology

Equations (4.44), (4.45), and (4.48) show the analytical relationship that relates the linear operator and the Gaussian noise in Eq. (4.30) to the statistical moments of the states of Eq. (4.38) that can be obtained through multiple experimental measurements or first-principles microscopic simulations.

To this end, we can follow a systematic procedure, similar to the one developed for the 1D case, to construct 2D linear stochastic PDEs for deposition processes on a 2D lattice. The procedure includes the following steps: First, we design a set of simulation experiments that cover the complete range of process operations; second, we run multiple simulations for each simulation condition to obtain the trajectories of the first and second statistical moments of the states (i.e., Fourier coefficients) computed from the surface snapshots; third, we compute the eigenvalues of the linear operator and covariance of the Gaussian noise based on the trajectories of the statistical moments of the states for each simulation condition, and determine the model parameters of the stochastic PDE (i.e., the prederivative coefficients and the order of the stochastic PDE); finally, we investigate the dependence of the model parameters of the stochastic PDE on the process parameters and determine the least-squares optimal form of the stochastic PDE model with model parameters expressed as functions of the process parameters. This procedure will be demonstrated using a thin-film deposition process described in Section 4.3.5 and can be readily extended to other stochastic processes.

Remark 4.3. We note that all the simulation experiments are executed using simulation lattices whose sizes are large enough to capture the dynamics of the surface evolution during the thin-film growth, and we run additional simulation experiments using larger lattices to check our results. As we have mentioned in Section 4.2, the experimental measurements obtained from the actual physical process can also be used for model construction, as long as the measurement has enough resolution to capture the surface evolution dynamics. Furthermore, since the same dynamics can be present at both large and small length scales, the resolution of the constructed model could be even better than the experimental measurements. In general, the use of a finite lattice size simulation and limited resolution measurements does not affect the accuracy of the constructed model, provided that the surface evolution dynamics are captured by the simulation/experimental data.

4.3.5 Application to a 2D Thin-Film Growth Process

In this section, we consider a thin-film growth process of deposition from the vapor phase in which the formation of the thin film is governed by three microscopic processes that occur on the surface, i.e., the adsorption of vapor-phase molecules on the surface, the migration of surface molecules, and the desorption of surface molecules. More specifically, we consider a single-species growth on a 2D lattice. The adsorption rate, which depends on the vapor-phase concentration, is considered uniform over the spatial domain. All surface sites are available for adsorption for all time and the adsorption rate for each surface site is denoted as W (expressed in number of molecules adsorbed per second, $1/s$).

The migration rate of each surface molecule depends on its local environment. Under the consideration of only first nearest-neighbor interactions, the migration rate of surface molecules from a surface site with n first nearest neighbors is given by

$$w_m(n) = k_{m0}e^{-\frac{E_s+nE_n}{k_BT}}, \tag{4.49}$$

where E_s is the energy barrier associated with migration due to surface effects, E_n is the energy barrier associated with migration due to nearest-neighbor interactions, k_{m0} is the frequency constant associated with migration, k_B is Boltzmann's constant, and T is the substrate temperature. The values of migration energy barriers and frequency constant used in this study are taken from the literature [133] for a molecular-beam epitaxy GaAs process and are as follows: $E_s = 1.58\,\text{eV}$, $E_n = 0.28\,\text{eV}$, and $k_{m0} = 2k_BT/h$, where h is Planck's constant.

The desorption rate of each surface molecule also depends on its local environment. Under the consideration of only first nearest-neighbor interactions, the desorption rate of surface molecules from a surface site with n first nearest neighbors is given by

$$w_d(n) = k_{d0}e^{-\frac{E_d+nE_n}{k_BT}}, \tag{4.50}$$

where E_d is the energy barrier associated with desorption due to surface binding, and k_{d0} is the frequency constant associated with desorption. A kMC simulation code is used to simulate the deposition process, and we use the values $E_d = 1.8\,\text{eV}$ and $k_{d0} = 2k_BT/h$ in all kMC simulations in this chapter. We note that in contrast to the 1D process of Section 4.2, which only includes adsorption and surface migration processes, the 2D process includes adsorption, surface migration, and desorption processes.

Eigenvalues and Covariance

The eigenvalues and the covariance of the systems of ODEs, which correspond to the deposition processes with different W and T values, are identified based on the trajectories of the statistical moments. In the previous subsection, we showed that for a deposition process with a flat initial surface, the trajectory of the second statistical moment of the ODE state $\langle z_{m,n}(t)z^*_{m,n}(t)\rangle$ can be predicted by Eq. (4.44). Therefore, we can fit ς^2 and $\text{Re}(\lambda_{m,n})$ in Eq. (4.44) for the profile of $\langle z_{m,n}(t)z^*_{m,n}(t)\rangle$. Similarly, $\text{Im}(\lambda_{m,n})$ can be determined based on the trajectory of $\langle \text{Re}[z_{m,n}(t)]^2\rangle$ and Eq. (4.48).

In order to obtain the profile of $\langle z_{m,n}(t)z^*_{m,n}(t)\rangle$ and $\langle \text{Re}[z_{m,n}(t)]^2\rangle$, we need to generate snapshots of the thin-film surface during each deposition simulation and compute the values of $z_{m,n}(t)$. Since the lattice consists of discrete sites, we let $h(k_xL, k_yL, t)$ be the height profile of the surface at time t with lattice constant L (k_x and k_y denote the coordinates of a specific surface site), and compute $z_{m,n}(t)$ as follows:

$$
\begin{aligned}
z_{m,n}(t) &= \int_0^\pi \int_0^\pi h(x,y,t)\phi^*_{m,n}(x,y)dxdy \\
&= \sum_{k_x,k_y=0}^{k_{\max}} h(k_xL, k_yL, t) \int_{k_xL}^{(k_x+1)L} \int_{k_yL}^{(k_y+1)L} \phi^*_{m,n}(x,y)dxdy,
\end{aligned}
\tag{4.51}
$$

where $k_{\max}L = \pi$ (i.e., the lattice is mapped to the domain $[0,\pi]^2$). Substituting Eq. (4.36) into Eq. (4.51), we can derive the following expressions for $z_{m,n}(t)$, $z_{0,n}(t)$, $z_{m,0}(t)$, and $z_{0,0}(t)$:

$$
\begin{aligned}
z_{m,n}(t) &= \sum_{k_x,k_y=0}^{k_{\max}} \frac{h(k_xL, k_yL, t)}{-4\pi mn} e^{-(I2mk_xL + I2nk_yL)} \\
&\quad \times (e^{-I2mL} - 1)(e^{-I2nL} - 1), \quad m,n = \pm 1,\ldots,\pm\infty,
\end{aligned}
\tag{4.52}
$$

$$
\begin{aligned}
z_{0,n}(t) &= \sum_{k_x,k_y=0}^{k_{\max}} \frac{h(k_xL, k_yL, t)Le^{-I2k_yLn}}{-I2\pi n}(e^{-I2Ln} - 1) \\
&\quad n = \pm 1,\ldots,\pm\infty,
\end{aligned}
\tag{4.53}
$$

$$z_{m,0}(t) = \sum_{k_x,k_y=0}^{k_{\max}} \frac{h(k_xL, k_yL, t)Le^{-I2k_xLm}}{-I2\pi m}(e^{-I2Lm} - 1) \quad (4.54)$$

$$m = \pm 1, \ldots, \pm\infty,$$

$$z_{0,0}(t) = \sum_{k_x,k_y=0}^{k_{\max}} \frac{h(k_xL, k_yL, t)L^2}{\pi}. \quad (4.55)$$

We note that, for each simulation experiment, the profile "s" of $\langle z_{m,n}(t)z_{m,n}^*(t)\rangle$ and $\langle \mathrm{Re}[z_{m,n}(t)]^2 \rangle$ are computed based on 100 simulation runs taking place with the same process parameters (a further increase in the number of simulations led to identical results for the order and the parameters of the constructed stochastic PDE).

Figure 4.8 shows an eigenspectrum computed from a thin-film deposition (we note that the computed eigenvalues are considered real since the imaginary part of the eigenvalues turned out to be very small). It can be seen that the computed spectrum is very close to the parabolic reference curve (appears as a line when the eigenvalue is plotted against $m^2 + n^2$). Based on Eq. (4.36), this implies that a second-order stochastic PDE system of the following form would be able to describe the evolution of the surface height of this deposition process:

$$\frac{\partial h}{\partial t} = c + c_2 \nabla^2 h + \xi(x, y, t),$$

$$\nabla h(0, y, t) = \nabla h(\pi, y, t), \quad h(0, y, t) = h(\pi, y, t),$$
$$\nabla h(x, 0, t) = \nabla h(x, \pi, t), \quad h(x, 0, t) = h(x, \pi, t), \quad (4.56)$$
$$h(x, y, 0) = h_0(x, y),$$

where c, c_2, and the covariance of the Gaussian noise ξ, ς all depend on the microscopic processes and operating conditions.

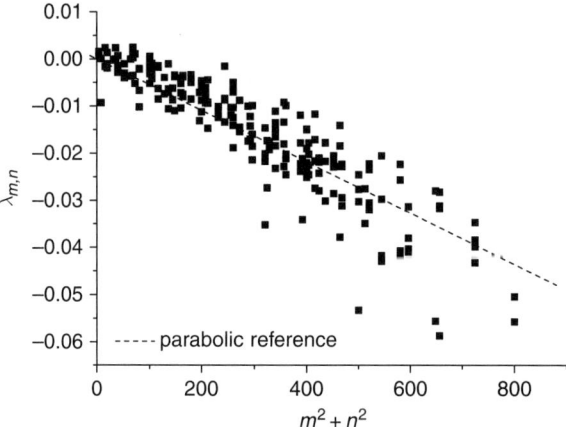

Fig. 4.8. Eigenvalue spectrum of the stochastic ODE systems computed from the kMC simulation of the deposition process with $W = 0.5\,\mathrm{s}^{-1}$, $T = 650\,\mathrm{K}$, and $k_{\max} = 100$.

Remark 4.4. We note that it is necessary to rescale $m^2 + n^2$ with the square of the corresponding lattice size, to carry out a meaningful comparison among eigenspectrums computed from simulations using lattices of different sizes; for the same reason, the covariance values should be scaled with the inverse of the square of the lattice size, $1/k_{max}^2$ (see Section 4.2 for a detailed discussion).

Dependence on the Process Parameters

We proceed now with the derivation of the parameters of the stochastic PDE of Eq. (4.56). c, c_2, and ς^2 are evaluated for assorted deposition conditions, and the lattice size of 100×100 (i.e., $k_{max} = 100$) is used for all simulation runs in our study.

The model parameter c is determined using Eq. (4.45) based on the trajectory of $\langle z_{0,0}(t) \rangle$. Since $z_{0,0}(t)$ is, in fact, proportional to the average height [see Eq. (4.55], i.e., the thickness of the film), c should equal the adsorption rate W when there is no desorption of surface molecules (see the process studied in Section 4.2 for example). However, desorption of surface molecules is significant in the deposition process studied in this chapter, and thus, the actual value of c should be smaller than W. Therefore, to derive the expression for c, we plot the relative difference of c and W [i.e., $(W - c)/W$] against W and T [118]. We find that $\ln[(W - c)/W]$ has a quasi-linear relationship with both T and $\ln W$; thus, the following expression can be obtained for c as a function of T and W through least-squares fitting:

$$c(W, T) = W \left(1 - \frac{k_w}{W^{a_w} e^{-k_B T/E_w}} \right) \tag{4.57}$$

where $k_w = 3.3829 \times 10^{-12}$, $a_w = 0.6042$, and $E_w = 2.7 \times 10^{-3}\,\text{eV}$.

The value of c_2 is determined by least-squares fitting of Eq. (4.36) and the eigenspectrum computed from the simulation. Based on the profile of c_2 as a function of T and W, we find that $\ln c_2$ has a quasi-linear relationship with both T and $\ln W$. Thus, the following expression can be obtained for c_2 as a function of T and W through least-squares fitting:

$$c_2(W, T) = \frac{k_{c0}}{W^{a_c} e^{-k_B T/E_c}} = \frac{k_c}{k_{max}^2 W^{a_c} e^{-k_B T/E_c}}, \tag{4.58}$$

where $k_c = 1.0274 \times 10^{-13}$, $a_c = 0.1669$, and $E_c = 1.9 \times 10^{-3}\,\text{eV}$.

The value of ς^2 is obtained by averaging the ς^2 values determined using Eq. (4.44) based on the trajectories of the second statistical moments of the states. However, derivation of the expression of $\varsigma^2(T, W)$ is not as straightforward as the ones for c and c_2. Consider the normalized ς^2 value, $\varsigma^2/(\pi/k_{max})^2$, as a function of T for different W. It is found that $\varsigma^2/(\pi/k_{max})^2$ grows exponentially with W; therefore, we may assume $\varsigma^2/(\pi/k_{max})^2 = 1 + e^{a_{v0} + k_{v0}T}$ [118]. Values of a_{v0} and k_{v0} are determined by least-squares fitting for different W, and results suggest that a_{v0} and k_{v0} are linear functions of W. Thus, the following expression is obtained for ς^2 as a function of T and W:

$$\varsigma^2(W,T) = \frac{\pi^2}{k_{\max}^2} W [1 + e^{-a_v - k_v W + (a_t + k_t W)T}]$$

$$= \frac{\pi^2}{k_{\max}^2} W \left[1 + \frac{e^{(a_t + k_t W)T}}{e^{a_v + k_v W}} \right], \tag{4.59}$$

where $a_v = 15.55493$, $k_v = 20.64504$, $a_t = 0.02332$, and $k_t = 0.0261$.

Therefore, the linear stochastic PDE model computed for the deposition process is as follows:

$$\frac{\partial h}{\partial t} = W \left(1 - \frac{k_w}{W^{a_w} e^{-k_B T / E_w}} \right)$$

$$+ \left(\frac{k_c}{k_{\max}^2 W^{a_c} e^{-k_B T / E_c}} \right) \nabla^2 h + \xi(x, y, t), \tag{4.60}$$

$$\nabla h(0, y, t) = \nabla h(\pi, y, t), \quad h(0, y, t) = h(\pi, y, t),$$
$$\nabla h(x, 0, t) = \nabla h(x, \pi, t), \quad h(x, 0, t) = h(x, \pi, t),$$
$$h(x, y, 0) = h_0(x, y),$$

where $\langle \xi(x, y, t) \xi^*(x', y', t') \rangle = \pi^2 / k_{\max}^2 W \left[1 + \left(e^{(a_t + k_t W)T} / e^{a_v + k_v W} \right) \right]$ $\delta(x - x')\delta(y - y')\delta(t - t')$.

Validation of the 2D Stochastic PDE Model

We now proceed with the validation of the stochastic PDE model of the thin-film deposition process [Eq. (4.60)]. Validation experiments are conducted for a number of deposition conditions that have not been used for the model construction. We generate surface profiles using both the stochastic PDE model and the kinetic Monte Carlo simulation. Figure 4.9 shows the surface profile at the end of a deposition with substrate temperature $T = 610$ K, adsorption rate $W = 0.5\,\mathrm{s}^{-1}$, deposition duration of 200 s, and $k_{\max} = 100$; Fig. 4.10 shows the surface profile at the end of a deposition with substrate temperature $T = 710$ K, adsorption rate $W = 0.5\,\mathrm{s}^{-1}$, deposition duration of 200 s, and lattice size $k_{\max} = 100$. We can see that at both low and high substrate temperatures, the linear stochastic PDE model constructed for the deposition process is very consistent with the kinetic Monte Carlo simulation in terms of film thickness and surface morphology (such as surface island size distribution and aggregation). The only observable difference between the two surfaces is that the one generated by kMC simulation has finer structural details than the one generated by stochastic PDE simulation. Such a difference is caused by the fact that the surface height profile in the stochastic PDE model is a continuous approximation of the discrete lattice.

In addition, we generate expected surface roughness profiles using both the stochastic PDE model and the kinetic Monte Carlo simulation (average of 100 runs)

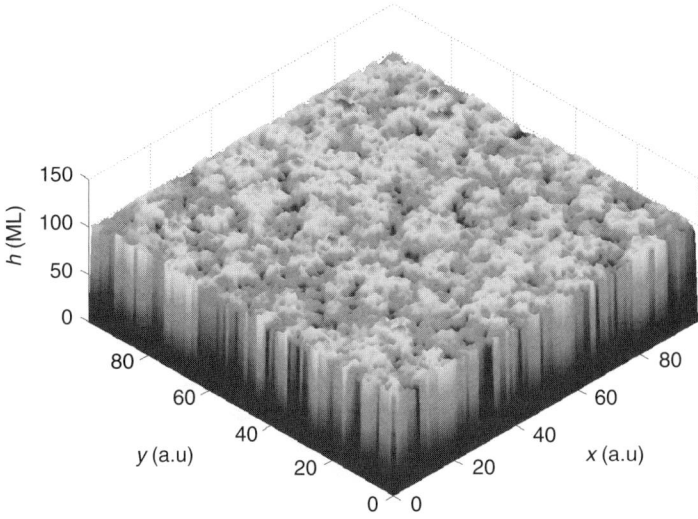

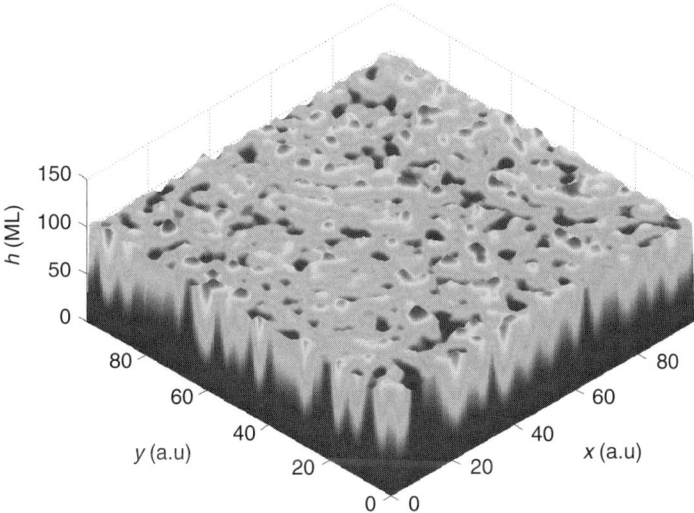

Fig. 4.9. Final thin-film surface profiles generated by kMC simulation (left, $k_{max} = 100$) and stochastic PDE model (right, 20×20 states) for a 200-s deposition with substrate temperature $T = 610\,\mathrm{K}$ and adsorption rate $W = 0.5\,\mathrm{s}^{-1}$.

for the deposition process. For simplicity, the surface roughness is evaluated in a root-mean-square fashion as follows:

$$r(t) = \sqrt{\frac{1}{\pi} \int_0^\pi \int_0^\pi [h(x, y, t) - \bar{h}(t)]^2 dx dy}, \qquad (4.61)$$

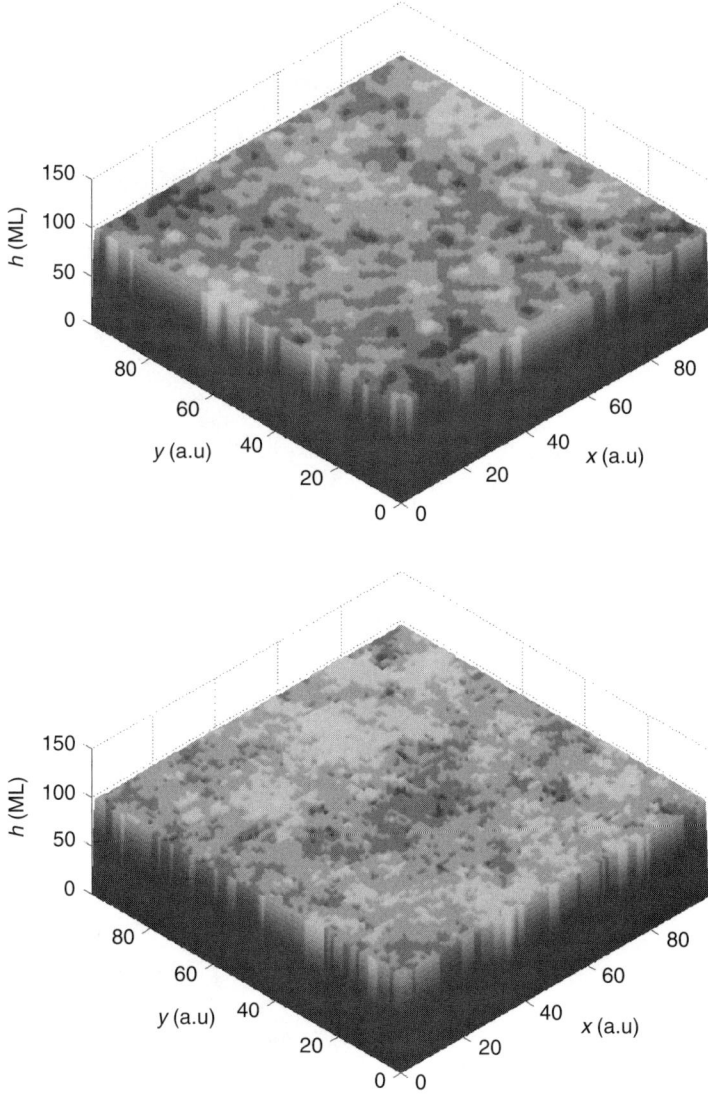

Fig. 4.10. Final thin-film surface profiles generated by kMC simulation (left, $k_{\max} = 100$) and stochastic PDE model (right, 20×20 states) for a 200 s deposition with substrate temperature $T = 710$ K and adsorption rate $W - 0.5s^{-1}$.

where $\bar{h}(t) = 1/\pi^2 \int_0^\pi \int_0^\pi h(x,y,t)dxdy$ is the average surface height. We note that for more detailed descriptions of the surface morphology, the surface can be examined using the height–height correlation function [144] and the interface width function [3]. Note that these different descriptions can be very efficiently computed using standard algebraic operations when the surface morphology is available from either the kMC model or the stochastic PDE model. The computational time required to obtain descriptions of the surface morphology is not significant compared to that needed for the solution of the surface microstructure model (either kMC model or stochastic PDE model).

To calculate the expected surface roughness using the stochastic PDE model, we first express the surface roughness in terms of the ODE states. According to Eq. (4.51), we have $\bar{h}(t) = z_{0,0}(t)\phi_{0,0}$. Therefore, $r(t)$ can be rewritten in terms of $z_{m,n}$ as follows:

$$
\begin{aligned}
r(t) &= \sqrt{\frac{1}{\pi^2} \int_0^\pi \int_0^\pi [h(x,y,t) - \bar{h}(t)][h(x,y,t) - \bar{h}(t)]^* dx dy} \\
&= \sqrt{\frac{1}{\pi^2} \int_0^\pi \int_0^\pi \sum_{\substack{m,n=-\infty, \\ m^2+n^2 \neq 0}}^{\infty} z_{m,n}(t)\phi_{m,n}(x,y)\phi_{m,n}^*(x,y)z_{m,n}^*(t) dx dy} \\
&= \sqrt{\frac{1}{\pi^2} \sum_{\substack{m,n=-\infty, \\ m^2+n^2 \neq 0}}^{\infty} z_{m,n}(t)z_{m,n}^*(t)},
\end{aligned}
\tag{4.62}
$$

and the expected roughness can be computed as follows:

$$
\langle r^2(t) \rangle = \frac{1}{\pi^2} \sum_{\substack{m,n=-\infty, \\ m^2+n^2 \neq 0}}^{\infty} \langle z_{m,n}(t)z_{m,n}^*(t) \rangle.
\tag{4.63}
$$

Substituting Eq. (4.42) and $\lambda_n = -4c_2(m^2 + n^2)$ into Eq. (4.63), we obtain the following expression of the trajectory of $\langle r(t) \rangle$ in terms of the parameters of the stochastic PDE model:

$$
\begin{aligned}
\langle r^2(t) \rangle = \frac{1}{\pi^2} \sum_{\substack{m,n=-\infty, \\ m^2+n^2 \neq 0}}^{\infty} \Bigg[&\varsigma^2 \frac{e^{-8c_2(m^2+n^2)t} - 1}{-8c_2(m^2+n^2)} \\
&+ e^{-8c_2(m^2+n^2)t} z_{m,n,0} z_{m,n,0}^* \Bigg].
\end{aligned}
\tag{4.64}
$$

Figure 4.11 shows the expected roughness profile of a deposition with adsorption rate $W = 0.5\,\text{s}^{-1}$ and different substrate temperatures: $T = 610\,\text{K}$ and $T = 710\,\text{K}$. We can see that the roughness profiles generated by the linear stochastic PDE model are very close to the profiles generated by the kinetic Monte Carlo simulation, for both low and high substrate temperatures.

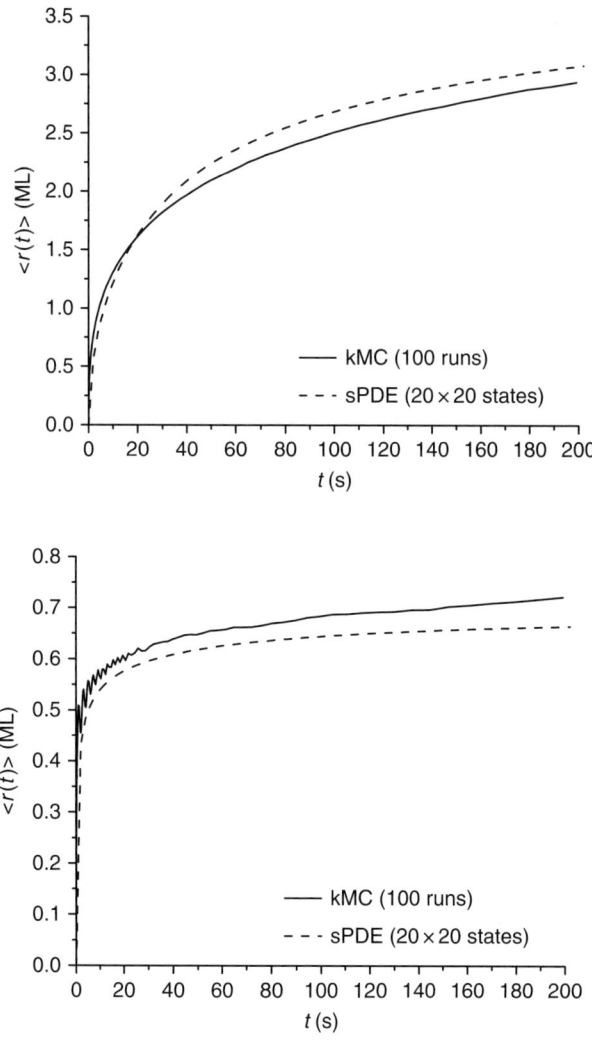

Fig. 4.11. Expected surface roughness profiles generated by kMC simulation ($k_{max} = 100$) and stochastic PDE model for a $200\,\mathrm{s}$ deposition with adsorption rate $W = 0.5\,\mathrm{s}^{-1}$ and different substrate temperatures: $T = 610\,\mathrm{K}$ (top) and $T = 710\,\mathrm{K}$ (bottom).

Furthermore, we have also generated expected thin-film thickness profiles using both the stochastic PDE model and the kinetic Monte Carlo simulation (average of 100 runs) for the deposition process [117]. Numerical simulations have demonstrated that the thickness profiles generated by the linear stochastic PDE model are also very close to the profiles generated by the kMC simulation, for both low and high substrate temperatures.

4.4 Parameter Estimation for Nonlinear Stochastic PDEs

In Sections 4.2 and 4.3, we presented methods for the construction of 1D and 2D linear stochastic PDE models for thin-film deposition processes. However, nonlinearities exist in many material preparation processes in which the surface evolution can be modeled by stochastic PDEs. A typical example of such processes is the sputtering process, whose surface evolution is described by the nonlinear stochastic Kuramoto–Sivashinsky equation (KSE). In a simplified setting, the sputtering process includes two types of surface micro–processes, erosion and diffusion. The nonlinearity of the sputtering process originates from the dependence of the erosion rate on a nonlinear sputtering yield function [36]. The presented methods for the identification and construction of linear stochastic PDEs require the analytical solutions for statistical moments, which prevent their direct applications to nonlinear stochastic PDEs.

Motivated by this section focuses on estimation of the parameters of nonlinear stochastic PDEs. To present the method we develop for parameter estimation, we use the nonlinear stochastic Kuramoto–Sivashinsky equation, a fourth-order nonlinear stochastic PDE, as a representative example of nonlinear stochastic PDEs. To perform this model parameter estimation task, we initially formulate the nonlinear stochastic KSE into a system of infinite nonlinear stochastic ODEs. A finite-dimensional approximation of the stochastic KSE is then constructed that captures the dominant mode contribution to the state and the evolution of the statistical moments of the state of the stochastic ODE system is derived. Then, a set of simulation experiments is designed that covers the complete range of process operations and multiple simulations are carried out for each simulation condition to obtain the trajectories of the statistical moments of the states computed from the surface snapshots. Finally, we determine the least-squares optimal form of the parameters of the nonlinear stochastic PDE model. The parameter estimation method is successfully applied to an ion-sputtering process that involves two surface microprocesses: atom erosion and surface diffusion.

4.4.1 Example: The Stochastic Kuramoto–Sivashinsky Equation

We consider the nonlinear stochastic Kuramoto–Sivashinsky equation (KSE), a fourth–order, nonlinear stochastic PDE [36], taking the following form:

$$\frac{\partial h}{\partial t} = -\nu \frac{\partial^2 h}{\partial x^2} - \kappa \frac{\partial^4 h}{\partial x^4} + \frac{\lambda}{2} \left(\frac{\partial h}{\partial x} \right)^2 + \xi(x, t) \qquad (4.65)$$

subject to periodic boundary conditions:

$$\frac{\partial^j h}{\partial x^j}(-\pi, t) = \frac{\partial^j h}{\partial x^j}(\pi, t), \qquad j = 0, \ldots, 3, \qquad (4.66)$$

with the initial condition

$$h(x,0) = h_0(x), \tag{4.67}$$

where ν, κ, and λ are parameters related to surface mechanisms [91], $x \in [-\pi, \pi]$ is the spatial coordinate, t is the time, and $h(x,t)$ is the height of the surface at position x and time t. The periodic boundary conditions are used so that the treatment of surface boundaries is consistent with that of the kMC model where periodic boundary conditions are also used. $\xi(x,t)$ is a Gaussian noise with the following expressions for its mean and covariance:

$$\langle \xi(x,t) \rangle = 0,$$

$$\langle \xi(x,t)\xi(x',t') \rangle = \sigma^2 \delta(x-x')\delta(t-t'), \tag{4.68}$$

where σ is a constant, $\delta(\cdot)$ is the Dirac function, and $\langle \cdot \rangle$ denotes the expected value. Note that the noise covariance depends on both space x and time t.

4.4.2 Model Reduction

To study the dynamics of Eq. (4.65), we initially consider the eigenvalue problem of the linear operator in Eq. (4.65), which takes the form

$$A\bar{\phi}_n(x) = -\nu \frac{d^2 \bar{\phi}_n(x)}{dx^2} - \kappa \frac{d^4 \bar{\phi}_n(x)}{dx^4} = \lambda_n \bar{\phi}_n(x),$$

$$\tag{4.69}$$

$$\frac{d^j \bar{\phi}_n}{dx^j}(-\pi) = \frac{d^j \bar{\phi}_n}{dx^j}(+\pi), \quad j = 0,\ldots,3, \quad n = 1,\ldots,\infty,$$

where λ_n denotes an eigenvalue and $\bar{\phi}_n$ denotes an eigenfunction. A direct computation of the solution of the above eigenvalue problem yields $\lambda_0 = 0$ with $\psi_0 = 1/\sqrt{2\pi}$, and $\lambda_n = \nu n^2 - \kappa n^4$ (λ_n is an eigenvalue of multiplicity two) with eigenfunctions $\phi_n = (1/\sqrt{\pi})\sin(nx)$ and $\psi_n = (1/\sqrt{\pi})\cos(nx)$ for $n = 1,\ldots,\infty$. Note that the $\bar{\phi}_n$ in Eq. (4.69) denotes either ϕ_n or ψ_n. From the expression of the eigenvalues, it follows that for fixed values of $\nu > 0$ and $\kappa > 0$, the number of unstable eigenvalues of the operator A in Eq. (4.69) is finite and the distance between two consecutive eigenvalues (i.e., λ_n and λ_{n+1}) increases as n increases.

To present the method that we use to estimate the parameters of the stochastic KSE in Eq. (4.65) and design controllers, we first derive a nonlinear stochastic ODE approximation of Eq. (4.65) using Galerkin's method. To this end, we first expand the solution in Eq. (4.65) in an infinite series in terms of the eigenfunctions of the operator of Eq. (4.69) as follows:

$$h(x,t) = \sum_{n=1}^{\infty} \alpha_n(t)\phi_n(x) + \sum_{n=0}^{\infty} \beta_n(t)\psi_n(x), \tag{4.70}$$

where $\alpha_n(t)$, $\beta_n(t)$ are time-varying coefficients. Substituting the above expansion for the solution, $h(x,t)$, into Eq. (4.65) and taking the inner product with the adjoint eigenfunctions, $\phi_n^*(z) = (1/\sqrt{\pi})\sin(nz)$ and $\psi_n^*(z) = (1/\sqrt{\pi})\cos(nz)$, the following system of infinite nonlinear stochastic ODEs is obtained:

$$\frac{d\alpha_n}{dt} = (\nu n^2 - \kappa n^4)\alpha_n + \lambda \cdot f_{n\alpha} + \xi_\alpha^n(t),$$

$$\frac{d\beta_n}{dt} = (\nu n^2 - \kappa n^4)\beta_n + \lambda \cdot f_{n\beta} + \xi_\beta^n(t)$$

(4.71)

For $n = 1, \ldots, \infty$, where

$$f_{n\alpha} = \frac{1}{2} \int_{-\pi}^{\pi} \phi_n^*(x) \cdot \left(\sum_{j=1}^{\infty} \alpha_j(t) \frac{d\phi_j}{dx}(x) + \sum_{j=0}^{\infty} \beta_j(t) \frac{d\psi_j}{dx}(x) \right)^2 dx,$$

(4.72)

$$f_{n\beta} = \frac{1}{2} \int_{-\pi}^{\pi} \psi_n^*(x) \cdot \left(\sum_{j=1}^{\infty} \alpha_j(t) \frac{d\phi_j}{dx}(x) + \sum_{j=0}^{\infty} \beta_j(t) \frac{d\psi_j}{dx}(x) \right)^2 dx,$$

and

$$\xi_\alpha^n(t) = \int_{-\pi}^{\pi} \xi(x,t) \phi_n^*(x) dx,$$

(4.73)

$$\xi_\beta^n(t) = \int_{-\pi}^{\pi} \xi(x,t) \psi_n^*(x) dx.$$

The covariances of $\xi_\alpha^n(t)$ and $\xi_\beta^n(t)$ can be computed by using Result 4.1 as follows: $\langle \xi_\alpha^n(t)\xi_\alpha^n(t')\rangle = \sigma^2\delta(t-t')$ and $\langle \xi_\beta^n(t)\xi_\beta^n(t')\rangle = \sigma^2\delta(t-t')$.

The surface roughness is represented by the standard deviation of the surface from its average height as defined in Eq. (3.2). According to Eq. (4.70), we have $\bar{h}(t) = \beta_0(t)\psi_0$. Therefore, $\langle r^2(t)\rangle$ can be rewritten in terms of $\alpha_n(t)$ and $\beta_n(t)$ as follows:

$$\langle r^2(t)\rangle = \frac{1}{2\pi} \left\langle \int_{-\pi}^{\pi} (h(x,t) - \bar{h}(t))^2 dx \right\rangle$$

$$= \frac{1}{2\pi} \left\langle \int_{-\pi}^{\pi} \left[\sum_{i=1}^{\infty} \alpha_i(t)\phi_i(x) + \sum_{i=0}^{\infty} \beta_i(t)\psi_i(x) - \beta_0(t)\psi_0 \right]^2 dx \right\rangle$$

(4.74)

$$= \frac{1}{2\pi} \left\langle \int_{-\pi}^{\pi} \sum_{i=1}^{\infty} [\alpha_i^2(t)\phi_i^2(x) + \beta_i^2(t)\psi_i^2(x)] dx \right\rangle$$

$$= \frac{1}{2\pi} \left\langle \sum_{i=1}^{\infty} (\alpha_i^2(t) + \beta_i^2(t)) \right\rangle = \frac{1}{2\pi} \sum_{i=1}^{\infty} [\langle \alpha_i^2(t)\rangle + \langle \beta_i^2(t)\rangle].$$

Eq. (4.74) provides a direct link between the state covariance of the infinite stochastic ODEs of Eq. (4.71) and the expected surface roughness of the sputtering process.

Due to its infinite-dimensional nature, the system in Eq. (4.71) cannot be directly used as a basis for either parameter estimation or feedback controller design that can be implemented in practice (i.e., the practical implementation of such algorithms will require the computation of infinite sums, which cannot be done by a computer). Instead, we will use finite-dimensional approximations of the system in Eq. (4.71).

Specifically, we rewrite the system in Eq. (4.71) as follows:

$$\frac{dx_s}{dt} = \Lambda_s x_s + \lambda \cdot f_s(x_s, x_f) + \xi_s,$$

$$\frac{dx_f}{dt} = \Lambda_f x_f + \lambda \cdot f_f(x_s, x_f) + \xi_f,$$

(4.75)

where

$$x_s = [\alpha_1 \quad \cdots \quad \alpha_m \quad \beta_1 \quad \cdots \quad \beta_m]^T,$$

$$x_f = [\alpha_{m+1} \quad \beta_{m+1} \quad \alpha_{m+2} \quad \beta_{m+2} \quad \cdots]^T,$$

$$\Lambda_s = \text{diag}[\lambda_1 \quad \cdots \quad \lambda_m \quad \lambda_1 \quad \cdots \quad \lambda_m],$$

$$\Lambda_f = \text{diag}[\lambda_{m+1} \quad \lambda_{m+1} \quad \lambda_{m+2} \quad \lambda_{m+2} \quad \cdots],$$

$$f_s(x_s, x_f) = [f_{1\alpha}(x_s, x_f) \quad \cdots \quad f_{m\alpha}(x_s, x_f) \quad f_{1\beta}(x_s, x_f) \quad \cdots$$

$$f_{m\beta}(x_s, x_f)]^T,$$

$$f_f(x_s, x_f) = [f_{(m+1)\alpha}(x_s, x_f) \quad f_{(m+1)\beta}(x_s, x_f)$$

$$f_{(m+2)\alpha}(x_s, x_f) \quad f_{(m+2)\beta}(x_s, x_f) \quad \cdots]^T,$$

$$\xi_s = [\xi_\alpha^1 \quad \cdots \quad \xi_\alpha^m \quad \xi_\beta^1 \quad \cdots \quad \xi_\beta^m]^T,$$

$$\xi_f = [\xi_\alpha^{m+1} \quad \xi_\beta^{m+1} \quad \xi_\alpha^{m+2} \quad \xi_\beta^{m+2} \quad \cdots]^T,$$

(4.76)

The dimension of the x_s subsystem is $2m$ and the x_f subsystem is infinite-dimensional.

Neglecting the x_f subsystem, the following $2m$-dimensional system is obtained:

$$\frac{d\tilde{x}_s}{dt} = \Lambda_s \tilde{x}_s + \lambda \cdot f_s(\tilde{x}_s, 0) + \xi_s,$$

(4.77)

where the tilde symbol in $\tilde{x}_s$ denotes that this state variable is associated with a finite-dimensional system.

4.4.3 System of Deterministic ODEs for State Covariance

The parameters of stochastic PDE models for many deposition and sputtering processes can be derived based on the corresponding master equation; they describe the evolution of the probability that the surface is at a certain configuration. For all practical purposes, the stochastic PDE model parameters should be estimated by matching the prediction of the stochastic PDE model to that of kMC simulations due to the approximations made in the derivation of the stochastic PDE model from the master equation [67, 101].

In this section, we present a method to estimate the parameters of the nonlinear stochastic KSE model of the sputtering process by using data from the kMC simulations of the process. The parameter-estimation algorithm is developed on the basis of the finite-dimensional system in Eq. (4.77).

The system in Eq. (4.77) is a finite-dimensional, nonlinear, stochastic ODE system including all four parameters, ν, κ, λ, and σ^2 of the stochastic KSE in Eq. (4.65). We first derive the system of deterministic ODEs that describes the dynamics of the covariance matrix of the state vector of Eq. (4.77), x_s, which is defined as $P_s = \langle x_s x_s^T \rangle$.

Consider the evolution of the state of Eq. (4.77) in a small time interval $[t, t+\Delta t]$ as follows [86, 35]:

$$x_s(t + \Delta t) = (I_s + \Delta t \cdot \Lambda_s) x_s(t) + \Delta t \cdot \lambda f_s(x_s(t), 0) + \Delta t \cdot \xi_s(t), \quad (4.78)$$

where I_s is a $2m \times 2m$ identity matrix. To study the dynamics of P_s, we approximate the Dirac function, $\delta(\cdot)$, involved in the covariances of ξ_s by $1/\Delta t$, and neglect the terms of order Δt^2. When Eq. (4.78) is used to compute the numerical solution of $x_s(t)$, it is clear that $x_s(t)$ is only dependent on $\xi_s(\tau)$ (for $\tau \leq t - \Delta t$). Since $\xi_s(t)$ and $\xi_s(\tau)$ are mutually independent according to the definition of Gaussian noise of Eq. (4.68) and Result 4.1, $\xi_s(t)$ is also independent of $x_s(t)$. We therefore have $\langle \xi_s(t) x_s^T(t) \rangle = 0$ and $\langle x_s(t) \xi_s^T(t) \rangle = 0$. Consequently, the following equation for P_s can be obtained from Eq. (4.78):

$$P_s(t + \Delta t) = P_s(t) + \Delta t \cdot \{\Lambda_s P_s(t) + P_s(t)\Lambda_s^T$$

$$+ \lambda \left\langle x_s(t) f_s(x_s(t), 0)^T + f_s(x_s(t), 0) x_s(t)^T \right\rangle + R_s\}, \quad (4.79)$$

where R_s is the intensity of ξ_s and $R_s \delta(t - t') = \langle \xi_s(t) \xi_s^T(t) \rangle$. In this work, $R_s = \sigma^2 I_{2m \times 2m}$.

By bringing $P_s(t)$ to the left-hand side of Eq. (4.79), dividing both sides by Δt, and setting $\Delta t \to 0$, we obtain the following nonlinear system of deterministic ODEs for the state covariance of the system in Eq. (4.75):

$$\frac{dP_s(t)}{dt} = \Lambda_s P_s(t) + P_s(t)\Lambda_s^T + R_s$$
$$+ \lambda \left\langle x_s(t) f_s(x_s(t), 0)^T + f_s(x_s(t), 0) x_s(t)^T \right\rangle. \quad (4.80)$$

Note that the linear part of Eq. (4.80) is the Lyapunov equation used in covariance controller design for linear systems [73]. We will use this deterministic ODE system as the basis for parameter estimation.

4.4.4 Parameter Estimation

The four parameters of the stochastic PDE process model in Eq. (4.65) can be estimated from Eq. (4.80). Specifically, the parameters ν and κ are included in the matrix Λ_s in Eq. (4.80) and the parameter λ is associated with the nonlinear term in Eq. (4.80). To this end, we need to obtain $P_s(t)$ and $\langle x_s(t) f_s(t)^T + f_s(t) x_s(t)^T \rangle$, which are both functions of x_s, to perform the parameter estimation.

The data of $x_s = [\alpha_1(t) \cdots \alpha_m(t) \beta_1(t) \cdots \beta_m(t)]^T$ can be obtained from kMC simulations of the sputtering process. Once x_s is obtained, $f_s(x_s, 0) = [f_{1\alpha}(x_s, 0) \cdots f_{m\alpha}(x_s, 0) f_{1\beta}(x_s, 0) \cdots f_{m\beta}(x_s, 0)]^T$ can be computed as follows:

$$f_{n\alpha}(x_s(t), 0) = \frac{1}{2} \int_{-\pi}^{\pi} \phi_n^*(x) \left(\sum_{j=1}^{m} \alpha_j(t) \frac{d\phi_j}{dx}(x) + \sum_{j=0}^{m} \beta_j(t) \frac{d\psi_j}{dx}(x) \right)^2 dx,$$

$$f_{n\beta}(x_s(t), 0) = \frac{1}{2} \int_{-\pi}^{\pi} \psi_n^*(x) \left(\sum_{j=1}^{m} \alpha_j(t) \frac{d\phi_j}{dx}(x) + \sum_{j=0}^{m} \beta_j(t) \frac{d\psi_j}{dx}(x) \right)^2 dx,$$

$$(4.81)$$

where $n = 1, 2, \cdots, m$. To compute the expected values for $x_s(t) \cdot x_s(t)^T$ and $x_s(t) f_s(x_s, 0)^T + f_s(x_s, 0) x_s(t)$, multiple kMC simulation runs for the sputtering process should be performed and the profiles of $x_s(t) \cdot x_s(t)^T$ and $x_s(t) f_s(x_s, 0)^T + f_s(x_s, 0) x_s(t)$ should be averaged to obtain the expected values.

The time derivative of $P_s(t)$ can be computed by the first-order approximation ($O(\Delta t)$) of the time derivative as follows:

$$\frac{dP_s(t)}{dt} = \frac{P_s(t + \Delta t) - P_s(t)}{\Delta t}, \qquad (4.82)$$

where Δt is a small time interval.

When the values of $dP_s(t)/dt$, $P_s(t)$, and $\langle x_s(t) f_s(x_s, 0)^T + f_s(x_s, 0) x_s(t)^T \rangle$ are obtained through kMC simulation runs at a set of discrete-time instants ($t = t_1, t_2, \ldots, t_k$), Eq. (4.80) becomes a system of linear algebraic equations for the four unknown model parameters. When the number of equations is larger than the number of parameters to be estimated, the least-squares method can be used to determine the model parameters.

Since P_s is a diagonally dominant matrix (see simulation part for a numerical verification), to make the parameter-estimation algorithm insensitive to round-off errors, we propose formulating the system of algebraic equations for least-squares fitting of the model parameters by using only the diagonal elements of the system in Eq. (4.80). The system of ODEs corresponding to the diagonal elements in Eq. (4.80) is as follows:

$$\frac{d\langle \alpha_n^2(t) \rangle}{dt} = 2(\nu n^2 - \kappa n^4) \cdot \langle \alpha_n^2(t) \rangle + 2\lambda \cdot \langle \alpha_n(t) \cdot f_{n\alpha}(t) \rangle + \sigma^2,$$

$$\text{(4.83)}$$

$$\frac{d\langle \beta_n^2(t) \rangle}{dt} = 2(\nu n^2 - \kappa n^4) \cdot \langle \beta_n^2(t) \rangle + 2\lambda \cdot \langle \beta_n(t) \cdot f_{n\beta}(t) \rangle + \sigma^2,$$

where $n = 1, \ldots, m$. The system in Eq. (4.83) is a linear system with respect to ν, κ, λ, and σ^2; reformulating Eq. (4.83) in the form of the following linear system to estimate ν, κ, λ and σ^2 using the least-squares method results to:

$$b = A\theta, \qquad \text{(4.84)}$$

where $\theta = [\nu \quad \kappa \quad \lambda \quad \sigma^2]^T$,

$$b = [b_1 \quad b_2 \quad \cdots \quad b_k \quad]^T,$$

$$\text{(4.85)}$$

$$b_i = \left[\frac{d\langle \alpha_1^2(t_i) \rangle}{dt} \quad \cdots \quad \frac{d\langle \alpha_m^2(t_i) \rangle}{dt} \quad \frac{d\langle \beta_1^2(t_i) \rangle}{dt} \quad \cdots \quad \frac{d\langle \beta_m^2(t_i) \rangle}{dt} \right]^T,$$

for $i = 1, 2, \ldots, k$,
and

$$A = \begin{bmatrix}
2 \cdot 1^2 \cdot \langle \alpha_1^2(t_1) \rangle & 2 \cdot 1^4 \cdot \langle \alpha_1^2(t_1) \rangle & 2\langle \alpha_1(t_1) \cdot f_{1\alpha}(t_1) \rangle & 1 \\
\vdots & \vdots & \vdots & \vdots \\
2 \cdot m^2 \cdot \langle \alpha_m^2(t_1) \rangle & 2 \cdot m^4 \cdot \langle \alpha_m^2(t_1) \rangle & 2\langle \alpha_m(t_1) \cdot f_{m\alpha}(t_1) \rangle & 1 \\
2 \cdot 1^2 \cdot \langle \beta_1^2(t_1) \rangle & 2 \cdot 1^4 \cdot \langle \beta_1^2(t_1) \rangle & 2\langle \beta_1(t_1) \cdot f_{1\beta}(t_1) \rangle & 1 \\
\vdots & \vdots & \vdots & \vdots \\
2 \cdot m^2 \cdot \langle \beta_m^2(t_1) \rangle & 2 \cdot m^4 \cdot \langle \beta_m^2(t_1) \rangle & 2\langle \beta_m(t_1) \cdot f_{m\beta}(t_1) \rangle & 1 \\
\vdots & \vdots & \vdots & \vdots \\
2 \cdot 1^2 \cdot \langle \alpha_1^2(t_k) \rangle & 2 \cdot 1^4 \cdot \langle \alpha_1^2(t_k) \rangle & 2\langle \alpha_1(t_k) \cdot f_{1\alpha}(t_k) \rangle & 1 \\
\vdots & \vdots & \vdots & \vdots \\
2 \cdot m^2 \cdot \langle \alpha_m^2(t_k) \rangle & 2 \cdot m^4 \cdot \langle \alpha_m^2(t_k) \rangle & 2\langle \alpha_m(t_k) \cdot f_{m\alpha}(t_k) \rangle & 1 \\
2 \cdot 1^2 \cdot \langle \beta_1^2(t_k) \rangle & 2 \cdot 1^4 \cdot \langle \beta_1^2(t_k) \rangle & 2\langle \beta_1(t_k) \cdot f_{1\beta}(t_k) \rangle & 1 \\
\vdots & \vdots & \vdots & \vdots \\
2 \cdot m^2 \cdot \langle \beta_m^2(t_k) \rangle & 2 \cdot m^4 \cdot \langle \beta_m^2(t_k) \rangle & 2\langle \beta_m(t_k) \cdot f_{m\beta}(t_k) \rangle & 1
\end{bmatrix}. \qquad \text{(4.86)}$$

Note that all elements in b and A can be obtained through the kMC simulations of the thin-film growth or sputtering process. The least-squares fitting of the model parameters can be obtained as follows:

$$\hat{\theta} = (A^T A)^{-1} A^T \cdot b. \qquad \text{(4.87)}$$

Remark 4.5. Note that it is important to appropriately collect the data set of surface snapshots from kMC simulations for parameter estimation. The data set should be representative so that the dynamics of the stochastic process can be adequately captured by the data set and reliable parameter estimation results can be obtained. Specifically, the condition number of the square matrix $A^T A$ in Eq. (4.87) should be used as an indicator of the quality of the data set. The matrix A is constructed by using the data derived from the surface snapshots. The condition number measures the sensitivity of the solution to the perturbations in A and b. There is stochastic noise contained in the data used to construct the matrix A and the vector b in Eq. (4.87). This noise will perturb A and b from their true values. A low condition number of the square matrix $A^T A$ will ensure that the perturbations in A and b introduced by the noise will not result in significant errors in the estimated model parameters. The sampling time and the number of surface snapshots should be carefully selected so that the condition number of the square matrix $A^T A$ is small.

4.4.5 Application to an Ion-Sputtering Process

In this section, we present applications of the presented model parameter-estimation method to the kMC model of a sputtering process to demonstrate the effectiveness of the algorithms. Specifically, the model parameters of the stochastic KSE process model are estimated using data of surface snapshots obtained from kMC simulations.

Description of the Sputtering Process

We consider a one-dimensional (1D) lattice representation of a crystalline surface of a sputtering process, which includes two surface microprocesses, atom erosion and surface diffusion. The solid-on-solid assumption is made, which means that no defects or overhangs are allowed to be developed in the film. The microscopic rules under which atom erosion and surface diffusion take place are as follows: A site, i, is first randomly picked among the sites of the whole lattice and the particle at the top of this site is subject to erosion with probability $0 < f < 1$, or diffusion with probability $1 - f$.

If the particle at the top of site i is subject to erosion, the particle is removed from site i with probability $P_e \cdot Y(\phi_i)$. P_e is determined as $1/7$ times the number of occupied sites in a 3×3 box centered at site i, which is shown in Fig. 4.12. There are nine sites in the box. The central one is the particle to be considered for erosion (the one marked by ●). Among the remaining eight sites, the site above the central site of interest must be vacant since the central site is a surface site. Therefore, only seven of the eight sites can be occupied and the maximum value of P_e is 1. $Y(\phi_i)$ is the sputtering yield function defined as follows:

$$Y(\phi_i) = y_0 + y_1 \phi_i^2 + y_2 \phi_i^4, \tag{4.88}$$

where y_0, y_1, and y_2 are constants. Following [36], the values of y_0, y_1, and y_2 can be chosen such that $Y(0) = 0.5$, $Y(\pi/2) = 0$ and $Y(1) = 1$, which correspond

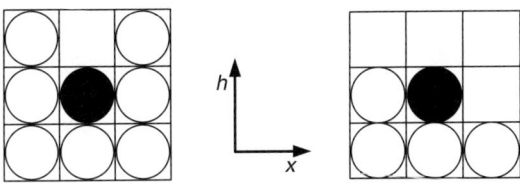

Fig. 4.12. Schematic of the rule to determine P_e. P_e is defined as $1/7$ times the number of occupied sites in a 3×3 box centered at the particle on the top of site i; $P_e = 1$ in the left figure and $P_e = 4/7$ in the right figure, where the particle marked by $\bullet$ is on the top of site i.

to $y_0 = 0.5$, $y_1 = 1.0065$, and $y_2 = -0.5065$. The local slope, ϕ_i, is defined as follows:

$$\phi_i = \tan^{-1}\left(\frac{h_{i+1} - h_{i-1}}{2a}\right), \tag{4.89}$$

where a is the lattice parameter and h_{i+1} and h_{i-1} are the values of the surface height at sites $i + 1$ and $i - 1$, respectively.

If the particle at the top of site i is subject to diffusion, one of its two nearest neighbors, j ($j = i+1$ or $i-1$), is randomly chosen and the particle is moved to the nearest-neighbor column with transition probability $w_{i \to j}$ as follows:

$$w_{i \to j} = \frac{1}{1 + \exp\left(\beta \Delta H_{i \to j}\right)}, \tag{4.90}$$

where $\Delta H_{i \to j}$ is the energy difference between the final and initial states of the move, $\beta = 1/k_B T$ and H is defined through the Hamiltonian of an unrestricted solid-on-solid model as follows:

$$H = \left(\frac{J}{a^n}\right) \sum_{k=1}^{N} (h_k - h_{k+1})^n, \tag{4.91}$$

where J is the bond energy, N is the total number of sites in the lattice, and n is a positive number. In the simulations presented here, we use $n = 2$ and $\beta J = 2.0$ [135]. Note that these microscopic rules have been used in a master equation approach to show that the continuum equation for the sputtering process is the stochastic KSE [91].

Kinetic Monte Carlo Simulation of the Sputtering Process

To carry out kMC simulations of this sputtering process, the rates of surface microprocesses should be computed [43, 155]. The rates of both erosion and diffusion are site-specific and can be obtained based on the process description as follows:

$$r_e(i) = \frac{f}{\tau} \cdot P_e(i) \cdot Y(\phi_i),$$

$$r_d(i, j) = \frac{1-f}{2\tau} \cdot w_{i \to j}; \qquad \text{for } i = 1, 2, \dots, N, \tag{4.92}$$

where $r_e(i)$ is the erosion rate at site i and $r_d(i, j)$ is the rate at which a surface particle hops from site i to site j. For the sputtering process considered, only nearest-neighbor hopping is allowed, so $j = i \pm 1$. $P_e(i)$ is determined by the box rule shown in Fig. 4.12, $Y(\phi_i)$ is defined in Eqs. (4.88) and (4.89), and $w_{i \to j}$ is defined in Eqs. (4.90) and (4.91). τ is defined as the time scale [91] and is fixed at 1/s for open-loop simulations in this work.

After the rates of surface microprocesses are determined, kMC simulations can be carried out using an appropriate algorithm. Specifically, we simulate the sputtering process in this work by using the null-event algorithm [162] so that the complex dependence of the transition probabilities on the surface microconfiguration in the sputtering process can be handled in an efficient way.

The following kMC simulation algorithm is used to simulate the sputtering process:

- The first integer random number, ζ_1 ($0 < \zeta_1 \leq N$, where ζ_1 is an integer and N is the total number of surface sites), is generated to pick a site i among all the sites on the 1D lattice.
- The second real random number, ζ_2 in the $(0, 1)$ interval, is generated to decide whether the chosen site i is subject to erosion ($\zeta_2 < f$) or diffusion ($\zeta_2 > f$).
- If the chosen site is subject to erosion, P_e and $Y(\phi_i)$ are computed. Specifically, P_e is computed by using the box rule shown in Fig. 4.12, where the center of the box is the surface particle on site i and $Y(\phi_i)$ is computed using Eqs. (4.88) and (4.89). Then, another real random number, ζ_{e3} in the $(0, 1)$ interval, is generated. If $\zeta_{e3} < P_e \cdot Y(\phi_i)$, the surface particle on site i is removed. Otherwise, no event is executed.
- If the chosen site is subject to diffusion, a side neighbor, j ($j = i + 1$ or $i - 1$ in the case of a 1D lattice) is randomly picked and the hopping rate, $w_{i \to j}$, is computed using Eq. (4.90). Then, another real random number, ζ_{d3} in the $(0, 1)$ interval, is generated. If $\zeta_{d3} < w_{i \to j}$, the surface atom is moved to the new site. Otherwise, no event is executed.
- Upon the execution of an event, a time increment, δt, is computed using the following expression:

$$\delta t = -\frac{\ln \zeta_4}{\frac{f}{\tau} \sum_{i=1}^{N} [P_e(i) \cdot Y(\phi_i)] + \frac{1-f}{2\tau} \sum_{i=1}^{N} [w_{i \to i+1} + w_{i \to i-1}]}, \quad (4.93)$$

where ζ_4 is a real random number in the $(0, 1)$ interval.

All random numbers, ζ_1, ζ_2, ζ_3, and ζ_4, follow a uniform probability distribution in their domains of definition.

Periodic boundary conditions (PBCs) are used in the kMC model of the sputtering process. Using PBCs, a particle that diffuses out of the simulation lattice at one boundary enters into the simulation lattice from the opposing side. Limited by the currently available computing power, the lattice size of a kMC simulation is much smaller than the size of a real process. Therefore, PBCs are widely used in microscopic simulations so that the statistical properties of a large-scale stochastic process can be appropriately captured by kMC simulations carried out on a small simulation lattice [107].

Simulation Results

In all simulations, we consider a sputtering process that takes place on a lattice containing 200 sites. Therefore, $a = 0.0314$. The sputtering yield function, $Y(\phi_i)$, is a nonlinear function of ϕ_i that takes the form of Eq. (4.88). y_0, y_1, and y_2 are chosen, such that $Y(0) = 0.5$, $Y(\pi/2) = 0$, and $Y(1) = 1$ [36].

We first compute the profiles of the state covariance and the expected values for $\alpha_n \cdot f_{n\alpha}$ and $\beta_n \cdot f_{n\beta}$ from kMC simulations of the sputtering process. Upon the execution of an event, the state of the stochastic KSE model (α_n or β_n) is updated. If the executed event is erosion, α_n or β_n can be updated as follows [101, 103]:

$$\alpha_n^{\text{new}} = \alpha_n^{\text{old}} + \frac{a\left[\psi(n, z_i - a/2) - \psi(n, z_i + a/2)\right]}{n},$$

$$\beta_n^{\text{new}} = \beta_n^{\text{old}} + \frac{a\left[\phi(n, z_i + a/2) - \phi(n, z_i - a/2)\right]}{n}.$$
(4.94)

If the executed event is diffusion from site i to site j, α_n or β_n are updated as follows:

$$\alpha_n^{\text{new}} = \alpha_n^{\text{old}} + \frac{a}{n} \cdot \{[\psi(n, z_i - a/2) - \psi(n, z_i + a/2)]$$

$$- [\psi(n, z_j - a/2) - \psi(n, z_j + a/2)]\},$$

$$\beta_n^{\text{new}} = \beta_n^{\text{old}} + \frac{a}{n} \cdot \{[\phi(n, z_i + a/2) - \phi(n, z_i - a/2)]$$

$$- [\phi(n, z_j + a/2) - \phi(n, z_j - a/2)]\},$$
(4.95)

where a is the lattice parameter and z_i is the coordinate of the center of site i.

The covariance profiles of α_1, α_3, α_5, α_7, and α_9 are shown in Fig. 4.13 and the profiles for the expected values of $\alpha_1 f_{1\alpha}$, $\alpha_3 f_{3\alpha}$, $\alpha_5 f_{5\alpha}$, $\alpha_7 f_{7\alpha}$, and $\alpha_9 f_{9\alpha}$ are shown in Fig. 4.14. Similar profiles are observed for the covariance of β_n and $\beta_n f_{n\beta}$ and are omitted here for brevity.

The terms $\alpha_n \cdot f_{n\alpha}$ and $\beta_n \cdot f_{n\beta}$ are computed using Eq. (4.81) with $m = 10$ for $n = 1, 2, \ldots, 10$. The expected profiles are the averages of profiles obtained from 10,000 independent kMC simulation runs. Since we use $m = 10$, the first $2m = 20$ modes are used for parameter estimation. The three-dimensional profile of the covariance matrix for the first 20 states at the end of a simulation run is plotted in Fig. 4.15. It is clear that the covariance matrix is diagonally dominant. Therefore, it is appropriate to use just the diagonal elements of the system in Eq. (4.80) for parameter estimation so that the estimation algorithm is insensitive to round-off errors.

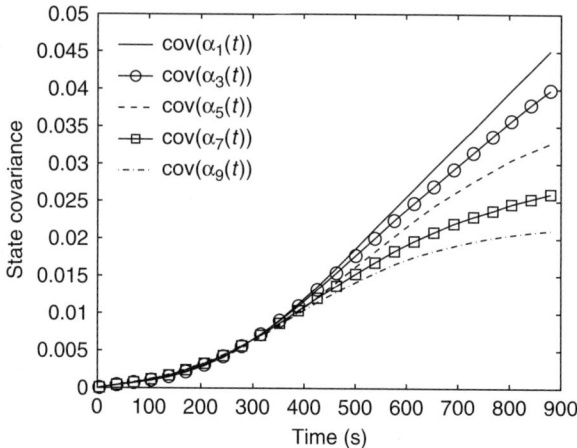

Fig. 4.13. Profiles of the state covariance $\langle \alpha_n^2(t) \rangle$ for $n = 1, 3, 5, 7,$ and 9.

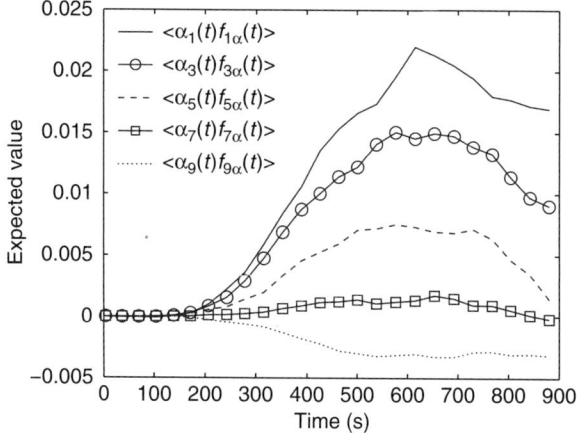

Fig. 4.14. Profiles of the expected value for $\alpha_n \cdot f_{n\alpha}(t)$ for $n = 1, 3, 5, 7,$ and 9.

To formulate the least-squares fitting problem, $d\langle \alpha_n^2(t) \rangle / dt$, $d\langle \beta_n^2(t) \rangle / dt$, $\langle \alpha_n^2(t) \rangle$, $\langle \beta_n^2(t) \rangle$, $\langle \alpha_n(t) \cdot f_{n\alpha}(t) \rangle$, and $\langle \beta_n(t) \cdot f_{n\beta}(t) \rangle$ are evaluated at the first 150 available discrete-time instants in the data obtained from kMC simulations. Therefore, in the least-squares fitting formulations of Eqs. (4.84) and (4.87), A is a $3,000 \times 4$ matrix, b is a $3,000 \times 1$ vector, and $\theta = [\nu \; \kappa \; \lambda \; \sigma^2]^T$. The values of the four parameters obtained from least-squares fitting are $\nu = 2.76 \times 10^{-5}$, $\kappa = 1.54 \times 10^{-7}$, $\lambda = 3.06 \times 10^{-3}$, and $\sigma^2 = 1.78 \times 10^{-5}$.

To validate the parameter-estimation method, we first compute the expected open-loop surface roughness from the stochastic KSE model in Eq. (4.65) with the computed parameters. A 200th-order stochastic ordinary differential equation approximation of the system in Eq. (4.65) is used to simulate the process (the use of

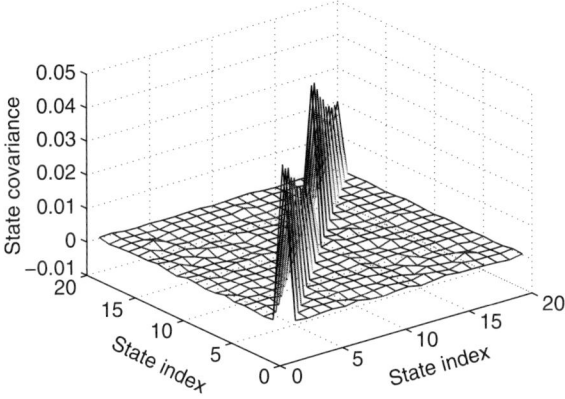

Fig. 4.15. The covariance matrix for the first 20 states: a diagonally dominant matrix.

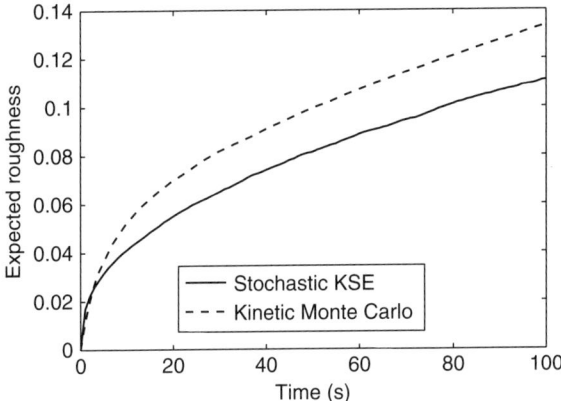

Fig. 4.16. Comparison of the open-loop profile of the expected surface roughness of the sputtering process from the kMC simulator and that from the solution of the stochastic KSE using the estimated parameters.

higher-order approximations led to identical numerical results, thereby implying that the following simulation runs are independent of the discretization). Then, the profile from the stochastic KSE with the computed parameters is compared to that from the kMC model. The expected surface roughness is computed from the simulations of the stochastic KSE and the kMC model by averaging surface roughness profiles obtained from 100 and 10,000 independent runs, respectively. The simulation result is shown in Fig. 4.16. It is clear that the computed model parameters result in consistent expected surface roughness profiles from the stochastic KSE model in Eq. (4.65) and from the kMC simulator of the sputtering process. There is an observable difference between the two profiles, which indicates the existence of a slight mismatch of the computed model with the kMC model of the sputtering process.

4.5 Conclusions

In this chapter, we have presented systematic methods for the construction of both linear and nonlinear stochastic PDE models for material microstructure using microscopic simulation data. In our development, a linear/nonlinear stochastic PDE was initially reformulated into a system of infinite stochastic ODEs by using modal decomposition. The statistical moments of the ODE states were consequently derived. For linear stochastic PDEs, the analytical solutions for the first and second statistical moments can be obtained, while for nonlinear stochastic PDEs, a system of deterministic nonlinear ODEs can be derived to capture the evolution of the statistical moments. Subsequently, kMC simulators of the stochastic processes were used to generate surface snapshots at different time instants during the process evolution to obtain the state and statistical moments of the stochastic ODE systems. Finally, the model parameters of the linear/nonlinear stochastic PDEs were obtained using least-squares fitting. The effectiveness of the proposed methods was demonstrated through applications to thin-film deposition processes taking place on both 1D and 2D lattices that are modeled by linear stochastic PDEs and to an ion-sputtering process that is modeled by a nonlinear stochastic PDE.

5

Feedback Control Using Stochastic PDEs

5.1 Introduction

Using the methods presented in Chapter 4, closed-form process models can be constructed for a variety of material preparation processes. When closed-form process models, in the form of linear or nonlinear stochastic PDEs, are available, it is desirable to design feedback controllers on the basis of process models. This has motivated research on the feedback control of surface roughness based on stochastic PDE process models.

In this chapter, we present methods for model-based controller design based on stochastic PDEs to control the thin-film surface roughness. A method for multivariable model predictive control using linear stochastic PDEs is first presented. The method results in the design of a computationally efficient multivariable predictive control algorithm that is successfully applied to the kMC models of thin-film deposition processes taking place in both 1D and 2D lattices to regulate the thin-film thickness and surface roughness at desired levels. When spatially distributed sensing and actuation are available, the surface roughness can be regulated using the covariance control technique. A method for linear covariance controller design is also presented. This method involves reformulation of the linear stochastic PDE into a system of infinite linear stochastic ordinary differential equations by using modal decomposition, derivation of a finite-dimensional approximation that captures the dominant mode contribution to the surface roughness, and feedback controller design based on the finite-dimensional approximation. The method for linear covariance control of surface roughness is demonstrated through application to the kMC model of a 1D thin-film growth process whose surface height evolution can be described by the Edwards–Wilkinson equation.

However, nonlinearities exist in many material preparation processes in which surface evolution can be modeled by stochastic PDEs. A typical example of such processes is the sputtering process, whose surface evolution is described by the nonlinear stochastic Kuramoto–Sivashinsky equation (KSE). To perform feedback control design for nonlinear stochastic processes, i.e., provide good performance for a wide range of process initial conditions and operating conditions, it is desirable that

P.D. Christofides et al., *Control and Optimization of Multiscale Process Systems*, Control Engineering, DOI 10.1007/978-0-8176-4793-3_5, © Birkhäuser Boston, a part of Springer Science+Business Media, LLC 2009

a nonlinear process model is directly used as the basis for controller synthesis. Motivated by this, we present a method for nonlinear covariance controller design based on nonlinear stochastic PDEs. The stochastic KSE is used to present the developed method. A finite-dimensional approximation of the stochastic KSE is first derived that captures the dominant mode contribution to the surface roughness, and a nonlinear feedback controller is designed based on this finite-dimensional approximation to control the surface roughness. An analysis of the closed-loop nonlinear infinite-dimensional system is performed to characterize the closed-loop performance enforced by the nonlinear feedback controller in the closed-loop infinite-dimensional system. The proposed nonlinear controller is successfully applied to a high-order approximation of the stochastic KSE and the kMC model of a sputtering process. The results of this chapter were first presented in [101–103, 115, 117].

5.2 Predictive Control Using Stochastic PDEs

This section focuses on model-predictive controller (MPC) design using linear stochastic PDEs. The linear stochastic PDE model in Eq. (4.28) is used to present the method. The methodology can be applied to other material preparation processes described by linear stochastic PDEs. The methodology is applied to a 1D thin-film growth process to control the final surface roughness and is applied to a 2D thin-film growth process to design a multivariable model-predictive controller to control the thin-film surface roughness and thickness. Numerical simulation results for these two applications are shown in Sections 5.2.3 and 5.2.4, respectively.

5.2.1 Generic MPC Formulation

Surface Roughness and Statistical Moments

We first proceed with the analysis of the dynamics of the surface roughness based on the stochastic PDE model constructed (Eq. (4.28)) for the thin-film deposition process. The surface roughness, $r(t)$, is represented by the standard deviation of the surface from its average height as defined in Eq. (3.2). According to Eq. (4.22), we have $\bar{h}(t) = z_0(t)\phi_0$. Therefore, $r(t)$ can be rewritten in terms of z_n as follows:

$$
\begin{aligned}
r(t) &= \sqrt{\frac{1}{\pi} \int_0^\pi (h(x,t) - \bar{h}(t))^2 dx} \\
&= \sqrt{\frac{1}{\pi} \int_0^\pi \sum_{n=-\infty,n\neq 0}^{\infty} z_n(t)\phi_n(x)\phi_n^*(x)z_n^*(t)dx} \qquad (5.1) \\
&= \sqrt{\frac{1}{\pi} \sum_{n=-\infty,n\neq 0}^{\infty} z_n(t)z_n^*(t),}
\end{aligned}
$$

and the expected roughness can be computed as follows:

$$\langle r^2(t) \rangle = \frac{1}{\pi} \sum_{n=-\infty, n \neq 0}^{\infty} \langle z_n(t) z_n^*(t) \rangle. \tag{5.2}$$

In order to design a model-based feedback controller to control the surface roughness, we first derive the analytical expression for the trajectory of $\langle r(t) \rangle$. Substituting Eq. (4.14) into Eq. (5.2), we obtain the following expression for $\langle r(t) \rangle$ in terms of the eigenvalues of the system of infinite stochastic ODEs in Eq. (4.9):

$$\langle r^2(t) \rangle = \frac{1}{\pi} \sum_{n=-\infty, n \neq 0}^{\infty} \left[\varsigma^2 \frac{e^{(\lambda_n + \lambda_n^*)t} - 1}{\lambda_n + \lambda_n^*} + e^{(\lambda_n + \lambda_n^*)t} z_{n0} z_{n0}^* \right]$$

$$= \frac{1}{\pi} \sum_{n=-\infty, n \neq 0}^{\infty} \left[\varsigma^2 \frac{e^{2\mathrm{Re}(\lambda_n)t} - 1}{2\mathrm{Re}(\lambda_n)} + e^{2\mathrm{Re}(\lambda_n)t} z_{n0} z_{n0}^* \right]. \tag{5.3}$$

Predictive Control Formulation

Since the thin-film deposition is a batch process, the control objective is to control the final surface roughness of the thin film to a desired level at the end of each deposition run by explicitly accounting for the presence of constraints on the manipulated input [an important problem that cannot be addressed by classical (PID) control schemes]. Therefore, we use an optimization-based control (model-predictive control-type) problem formulation. Figure 5.1 shows the block diagram of the closed-loop system. When a real time surface profile measurement is obtained, the states of the infinite stochastic ODE system, z_n, are computed. Then, a substrate temperature T is computed based on states z_n and the stochastic PDE model and applied to the deposition process. The substrate is held at this temperature for the rest of the deposition until a different value is assigned by the controller. The value of T is determined at each time t by solving, in real-time, an optimization problem minimizing the difference between the estimated final surface roughness and the desired level for this variable.

5.2.2 Order Reduction

In order to compute an estimate of the expected surface roughness at a future time t, we need to compute the infinite sum in Eq. (5.3). However, such an infinite summation cannot be computed directly; instead, an mth-order approximation (only the first mth states are included in the summation) needs to be used to approximately compute this infinite sum.

As an example, when the stochastic PDE model in Eq. (4.25) is considered, $\lambda_n = -4c_2 n^2$. Thus, Eq. (5.3) can be rewritten as follows:

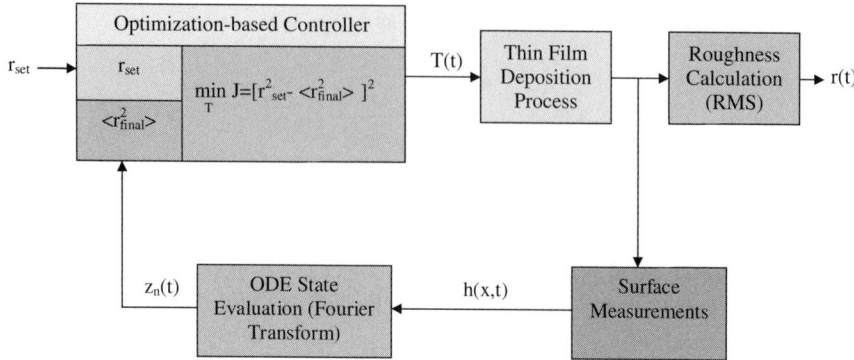

Fig. 5.1. Block diagram of the closed-loop system.

$$\langle r^2(t) \rangle = \frac{1}{\pi} \sum_{n=-\infty, n \neq 0}^{\infty} \left(\varsigma^2 \frac{e^{-8c_2 n^2 t} - 1}{-8c_2 n^2} + e^{-8c_2 n^2 t} z_{n0} z_{n0}^* \right)$$

(5.4)

$$= \frac{2}{\pi} \sum_{n=1}^{\infty} \left(\varsigma^2 \frac{e^{-8c_2 n^2 t} - 1}{-8c_2 n^2} + e^{-8c_2 n^2 t} z_{n0} z_{n0}^* \right).$$

Using the standard theory of infinite summation [89], it can be shown that, if the following mth-order approximation is used,

$$\hat{r}(t)^2 = \frac{2}{\pi} \sum_{n=1}^{m} \left(\varsigma^2 \frac{e^{-8n^2 c_2 t} - 1}{-8n^2 c_2} + e^{-8n^2 c_2 t} z_{n0} z_{n0}^* \right)$$

$$+ \frac{1}{2\pi} \left[e^{-8c_2(m+1)^2 t} \left(\pi r_0^2 - \sum_{n=1}^{m} z_{n0} z_{n0}^* \right) + \frac{\varsigma^2}{2^{m+2} c_2} \right],$$

(5.5)

where r_0 is the initial roughness value, the approximation error would be subject to the following bound:

$$|\langle r^2(t) \rangle - \hat{r}^2(t)| \leq \frac{1}{2\pi} \left[e^{-8c_2(m+1)^2 t} \left(\pi r_0^2 - \sum_{n=1}^{m} z_{n0} z_{n0}^* \right) + \frac{\varsigma^2}{2^{m+2} c_2} \right].$$

(5.6)

We note that the approximation error decreases with increasing m.

In general, to achieve a control precision ϵ, m should be chosen large enough for each optimization computation so that the approximation error is less than ϵ. However, to achieve the same control precision, the minimum m needed may vary depending on the specific surface configuration (i.e., current states z_n). On one hand, when the length scale of the surface fluctuation is very small, the magnitude of the high-order states becomes significant; hence, m needs to be relatively large so that these high-order states are included in the calculation. On the other hand, when the length scale of the surface fluctuation is relatively large, the contribution from the high-order states becomes negligible compared to the low-order states; hence, a relatively small m should be good enough for precise calculation. Therefore, in our implementation, the desired control precision is achieved by adding more states to the finite-dimensional system until the approximation error [computed based on Eq. (5.6)] is small enough, rather than by specifying the number of states that should be evaluated from the surface snapshot before hand. However, a limit on the maximum number of states to be used is imposed to guarantee that the computation time of the controller does not prevent real-time implementation (control precision may be reduced to meet real-time computation requirements).

5.2.3 Application to a 1D Thin-Film Growth Process

In this section, we design a state feedback model-predictive controller based on the stochastic PDE model in Eq. (4.28) to control the thin-film surface roughness of the 1D thin-film deposition process described in Section 4.2.5. Since the thin-film deposition is a batch process, the control objective is to control the final surface roughness of the thin film to a desired level at the end of each deposition run. Therefore, we use an optimization-based control problem formulation. The substrate temperature T is chosen to be the manipulated variable, while the thin-film growth rate W is kept constant during each deposition. Furthermore, since the process is stochastic, the controlled variable is the expected value of the final surface roughness, $\langle r^2(t_{\text{dep}}) \rangle$, where t_{dep} is the total deposition time.

The control system is designed based on the block diagram shown in Fig. 5.1. The value of T is determined at each time t by solving, in real time, the following optimization problem:

$$\min_{T} J = (r_{\text{set}}^2 - \langle r_{\text{final}}^2 \rangle)^2 \tag{5.7}$$

subject to

$$\langle r_{\text{final}}^2 \rangle = \frac{2}{\pi} \sum_{n=1}^{m} \left[\varsigma^2 \frac{e^{-8n^2 c_2(t_{\text{dep}} - t)} - 1}{-8n^2 c_2} + e^{-8n^2 c_2(t_{\text{dep}} - t)} z_n(t) z_n^*(t) \right]$$
$$+ \frac{1}{2\pi} \left\{ e^{-8c_2(m+1)^2(t_{\text{dep}} - t)} \left[\pi r^2(t) - \sum_{n=1}^{m} z_n(t) z_n^*(t) \right] + \frac{\varsigma^2}{2^{m+2} c_2} \right\}, \tag{5.8}$$

$$c_2 = \frac{e^{-32.002 + 0.0511T - 0.1620W}}{k_{\text{max}}^2}, \tag{5.9}$$

$$\varsigma^2 = \frac{\pi W}{k_{\text{max}}}, \tag{5.10}$$

$$T_{\min} \le T \le T_{\max}, \tag{5.11}$$

where $T_{\min}$ and $T_{\max}$ are the lowest and highest substrate temperature, respectively. We note that J corresponds to the difference between the square of the desired final surface roughness r_{set}^2 and the square of the estimated final surface roughness $\langle r_{\text{final}}^2 \rangle$ computed based on the current states z_n. We choose to minimize the difference of the squares of the surface roughness, i.e., the mean square of the surface height, to simplify the calculation.

The first equality constraint Eq. (5.8) [essentially the same as in Eq. (5.5)] states that the estimate of the final surface roughness, r_{final}^2, is computed based on current states $z_n(t)$ under the assumption that a substrate temperature T will be used and kept constant in the rest of the deposition. The second and third equality constraints are, in fact, Eqs. (4.26) and (4.27) of the stochastic PDE model of the deposition process, and since the growth rate W is fixed during each deposition, the third constraint can be removed by substituting the actual value of ς^2 into Eq. (5.8).

To solve the optimization problem in Eqs. (5.7)–(5.11), our initial step is to reduce it to a quadratic programming problem with only linear constraints. To do this, we remove the second and fourth constraints [Eqs. (5.9) and (5.11)] by first finding the optimal c_2 that minimizes J and then computing the corresponding optimal T using the equality constraint in Eq. (5.9). In addition, we linearize the first constraint in Eq. (5.8) with respect to c_2 around an initial guess $\tilde{c}_2$ (we note that when $\tilde{c}_2$ is chosen close enough to the optimal c_2, the solution of the linearized problem should be close to the solution of the original problem). The value of $\tilde{c}_2$ is computed based on the substrate temperature used in Eq. (5.9) (at $t = 0$, $\tilde{c}_2$ is computed based on the initial substrate temperature). Therefore, the original optimization problem is reduced to

$$\min_{c_2} J = (r_{\text{set}}^2 - \langle r_{\text{final}}^2 \rangle)^2 \tag{5.12}$$

subject to

$$\langle r_{\text{final}}^2 \rangle = \langle r_{\text{final}}^2(\tilde{c}_2) \rangle + (c_2 - \tilde{c}_2) \frac{\partial \langle r_{\text{final}}^2(\tilde{c}_2) \rangle}{\partial \tilde{c}_2}, \tag{5.13}$$

$$c_{2,\min} \le c_2 \le c_{2,\max}, \tag{5.14}$$

where $c_{2,\min}$ and $c_{2,\max}$ are the lower and upper bound of c_2, respectively. The second constraint is added due to the fact that c_2 can only take values within the corresponding range specified by Eqs. (5.9) and (5.11), and $c_{2,\min}$ and $c_{2,\max}$ are determined as follows:

$$c_{2,\min} = \frac{e^{-32.002 + 0.0511 T_{\min} - 0.1620W}}{k_{\max}^2},$$

$$c_{2,\max} = \frac{e^{-32.002 + 0.0511 T_{\max} - 0.1620W}}{k_{\max}^2}. \tag{5.15}$$

A standard procedure based on the active set method [56] is used to solve the optimization problem in Eq. (5.12). First, we drop the inequality constraint in Eq. (5.14), and thus, a direct computation of the above problem by substituting the equality constraint into the objective function yields

$$\bar{c}_2 = \tilde{c}_2 + \frac{r_{\text{set}}^2 - \langle r_{\text{final}}^2(\tilde{c}_2)\rangle}{\dfrac{\partial \langle r_{\text{final}}^2(\tilde{c}_2)\rangle}{\partial \tilde{c}_2}}, \tag{5.16}$$

where $\bar{c}_2$ is the optimal value of c_2 without the inequality constraint in Eq. (5.14). Then, we check whether the inequality constraint is violated by $\bar{c}_2$. If the inequality constraint is inactive (i.e., the constraint is not violated), $\bar{c}_2$ is considered to be the optimal value for the linearized optimization problem. On the other hand, if the inequality constraint is active (i.e., the constraint is violated), the optimization problem is resolved accounting for the active constraint (which serves as another equality constraint). In such a case, c_2 can only take the value of $c_{2,\text{min}}$ (when the lower bound is violated by $\bar{c}_2$) or $c_{2,\text{max}}$ (when the upper bound is violated by $\bar{c}_2$); hence, the optimal value is just the only feasible value, $c_{2,\text{min}}$ or $c_{2,\text{max}}$.

However, since Eq. (5.13) is the linearization of Eq. (5.8), $\bar{c}_2$ might only be a suboptimal value for the original problem. To this end, we can use this suboptimal $\bar{c}_2$ as a new guess and repeat the linearization procedure until $\bar{c}_2$ converges to the optimal value (the convergence is guaranteed if the original problem is convex), but for the sake of simplicity, such an iterative procedure is not adopted in this work. Once the optimal c_2 is determined, by substituting c_2 into Eq. (5.9), the optimal T can be obtained and used as the output of the controller.

Closed-Loop Simulation

A kMC simulation with a lattice size $k_{\text{max}} = 1000$ is used to simulate the thin-film deposition process, and the substrate temperature is restricted within 300 K to 900 K. The measurement interval, as well as the control interval, is set to be 1 s. We limit the maximum number of states to be used (in our case, to $m = 500$) to guarantee that the maximum possible computation time for each control action is within a certain practical requirement. Most of the time, however, the number of states needed by the controller is much smaller.

Figure 5.2 shows the surface roughness and substrate temperature profiles of a closed-loop deposition process with thin-film growth rate $W = 0.5\,\text{ML/s}$ and of an open-loop deposition with the same growth rate and a fixed substrate temperature $T = 650$ K. The control objective is to drive the final surface roughness of the thin film to 1.0 ML (monolayers) at the end of the 1000-s deposition. It can be seen that the final surface roughness is controlled at the desired level while an open-loop deposition with the same initial deposition condition would lead to a 100% higher final surface roughness, as shown in Fig. 5.2 (a comparison between the surfaces of the thin films deposited with closed-loop and open-loop deposition is shown in Fig. 5.3).

Figure 5.4 shows the final surface roughness histogram of the thin films deposited using 100 different closed-loop depositions with a final surface roughness set-point

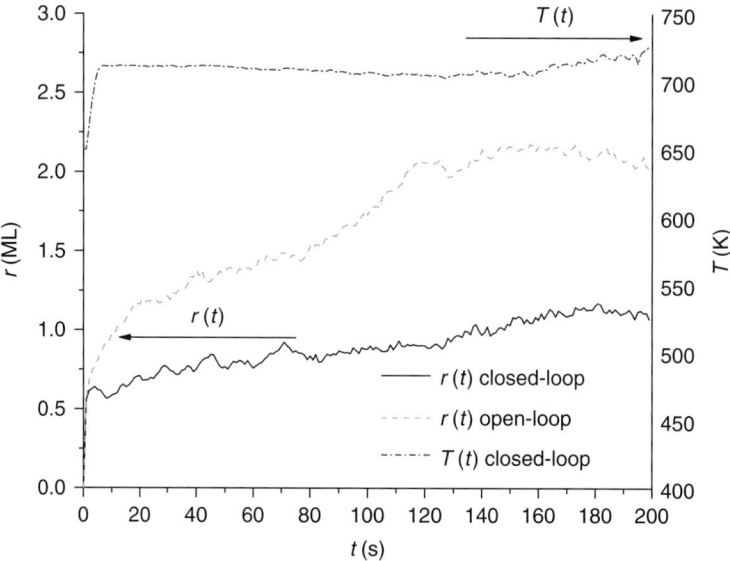

Fig. 5.2. Surface roughness and substrate temperature profiles of a 1000-s closed-loop deposition process with thin-film growth rate $W = 0.5$ ML/s and final roughness setpoint $r_{\text{set}} = 1.0$ ML.

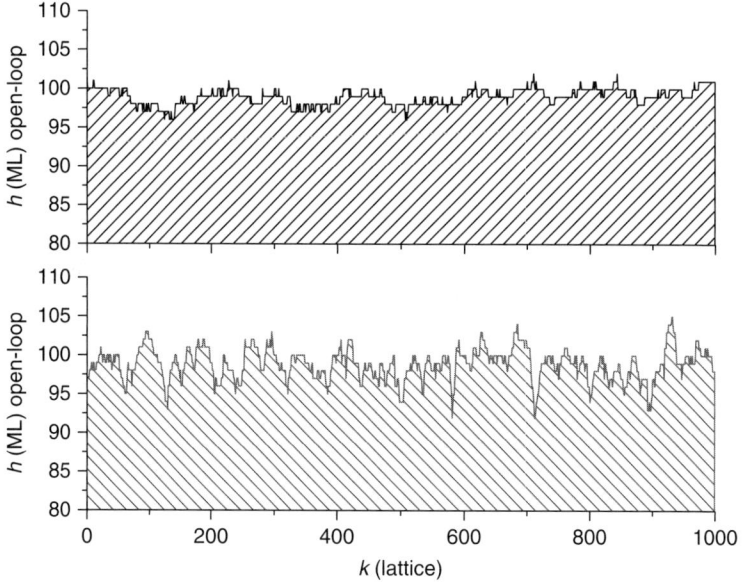

Fig. 5.3. Final thin-film surface profile of a closed-loop (top plot) and of an open-loop (bottom plot) deposition with the same initial substrate temperature $T = 650$ K, thin-film growth rate $W = 0.5$ ML/s, and deposition length $t_{\text{dep}} = 1000$ s .

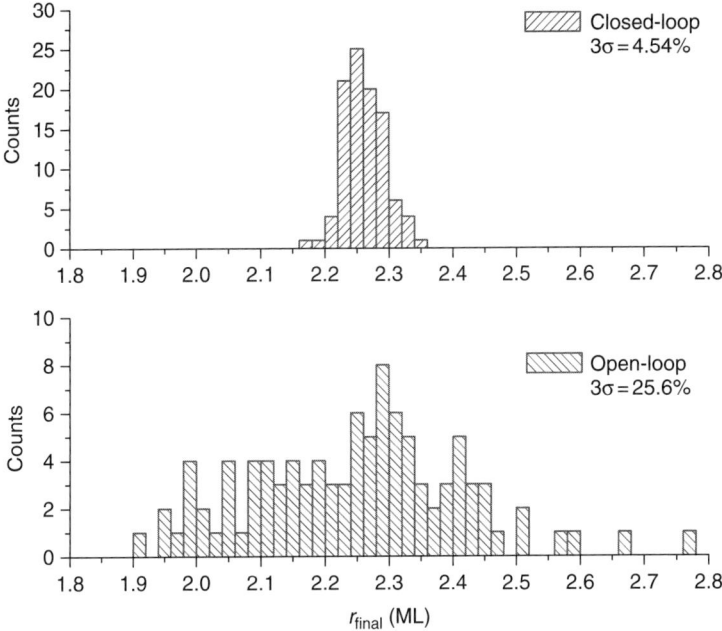

Fig. 5.4. Histogram of final surface roughness of 100 closed-loop and 100 open-loop thin-film depositions targeted at the same surface roughness level.

of 2.25 ML and 100 different open-loop depositions. It can be seen that the average surface roughness of the thin films deposited by the open-loop depositions is very close to the average surface roughness of the thin films deposited by the closed-loop deposition and the well-designed, recipe-based open-loop depositions. However, the variance among the thin films from different open-loop deposition runs is over 400% higher than that of closed-loop deposition runs even though no process disturbance is considered in the simulations. This is due to the fact that the stochastic nature of the microscopic processes of the film growth cannot be handled effectively without having a real-time feedback controller that can compensate for the stochastic variation from the expectation. As a result, if the tolerance on the thin-film surface roughness to fabricate a certain device is ± 0.1 ML, over half of the thin films prepared by the recipe-based, open-loop deposition would be disqualified. Therefore, introducing a real-time feedback control system that directly aims at the material and electrical properties of the thin films is one of the most effective, if not the only, solution to reduce cost and meet the ever-increasing film-quality requirements demanded by the devices, which are already down to the nanometer regime.

5.2.4 Application to a 2D Thin-Film Growth Process

In this section, we design a multivariable state feedback model-predictive controller based on the stochastic PDE model in Eq. (4.28) to control the thin-film thickness

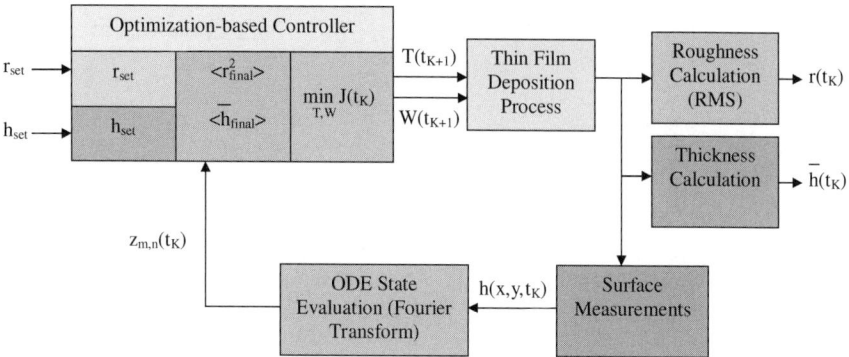

Fig. 5.5. Block diagram of the closed-loop system.

and surface roughness for the deposition process described in Section 4.3.5. The control objective is to control the final thin-film thickness and surface roughness to the desired levels at the end of each deposition run. We use an optimization-based control problem formulation. The substrate temperature T and the adsorption rate W (W can be adjusted by varying reactor inlet gas flow rate, chamber pumping speed, etc.) are chosen to be the manipulated variables. Furthermore, since the process is stochastic in nature, the controlled variables are the expected values of the final thin film thickness $\langle \bar{h}(t_{\text{dep}}) \rangle$ and of the surface roughness $\langle r^2(t_{\text{dep}}) \rangle$, where t_{dep} is the total deposition time.

Figure 5.5 shows the block diagram of the closed-loop system. The control system operates in a discrete-time (sample and hold) fashion: When the Kth real-time surface profile measurement is obtained at time t_K (i.e., $t_K = Kt_s$, where t_s is the measurement interval as well as the control interval), the states of the infinite stochastic ODE system, $z_{m,n}(t_K)$, are computed. Then, a substrate temperature $T(t_{K+1})$ and an adsorption rate $W(t_{K+1})$ are computed based on states $z_{m,n}(t_K)$ and the stochastic PDE model, under the assumption that T and W are held at designated levels for the rest of the deposition. $T(t_{K+1})$ and $W(t_{K+1})$ are then applied to the deposition process at the next measurement time, t_{K+1}.

Predictive Control Design

In order to design a model-based predictive controller, we first derive the analytical expression for the trajectory of $\langle \bar{h}(t) \rangle$ and $\langle r^2(t) \rangle$ for the 2D surface. Due to the fact that the current deposition parameters [$T(t_K)$ and $W(t_K)$] would be used during the current control cycle before the new levels [$T(t_{K+1})$ and $W(t_{K+1})$] are applied, the estimate of the film thickness (i.e., the estimate of $\langle z_{0,0}(t_{\text{dep}}) \rangle$) and the estimate of the final surface roughness cannot be computed directly using Eqs. (4.45) and (4.64) ($\lambda_{m,n}$ and ς^2 are no longer constant due to the change in W and T). Therefore, we first need to derive the expressions of $z_{0,0}(t_{\text{dep}})$ and $z_{m,n}(t_{\text{dep}})$ ($m^2 + n^2 \neq 0$)

for this case. We consider that at time t_{K+1}, the deposition parameters are changed from $W(t_K)$ and $T(t_K)$ to $W(t_{K+1})$ and $T(t_{K+1})$, respectively. Following from Eq. (4.40), we have

$$z_{0,0}(t) = z_{0,0}(t_0) + c_{0,0}^z(t - t_0) + \int_{t_0}^t \xi_{0,0}(\mu)d\mu,$$

$$z_{m,n}(t) = e^{\lambda_{m,n}(t - t_0)} z_{m,n}(t_0) + \int_{t_0}^t e^{\lambda_{m,n}(t - \mu)} \xi_{m,n}(\mu)d\mu, \tag{5.17}$$

$$m, n = 0, \pm 1, \ldots, \pm\infty, \qquad m^2 + n^2 \neq 0.$$

Hence, by calculating the intermediate values $z_{0,0}(t_{K+1})$ and $z_{m,n}(t_{K+1})$ $(m^2 + n^2 \neq 0)$ using $z_{0,0}(t_K)$ and $z_{m,n}(t_K)$, respectively, the expressions of $z_{0,0}(t_{\text{dep}})$ and $z_{m,n}(t_{\text{dep}})$ can be derived as follows:

$$z_{0,0}(t_{\text{dep}}) = z_{0,0}(t_K) + c_{0,0}^z(t_K)t_c + c_{0,0}^z(t_{K+1})(t_{\text{dep}} - t_{K+1})$$
$$+ \int_{t_K}^{t_{K+1}} \xi_{0,0}(\mu)d\mu + \int_{t_{K+1}}^{t_{\text{dep}}} \xi_{0,0}(\mu)d\mu,$$

$$z_{m,n}(t) = e^{\lambda_{m,n}(t_K)t_c + \lambda_{m,n}(t_{K+1})(t_{\text{dep}} - t_{K+1})} z_{m,n}(t_K)$$
$$+ e^{\lambda_{m,n}(t_{K+1})t_c} \int_{t_K}^{t_{K+1}} e^{\lambda_{m,n}(t_K)(t_{K+1} - \mu)} \xi_{m,n}(\mu)d\mu \tag{5.18}$$

$$+ \int_{t_{K+1}}^{t_{\text{dep}}} e^{\lambda_{m,n}(t_{K+1})(t_{\text{dep}} - \mu)} \xi_{m,n}(\mu)d\mu,$$

$$m, n = 0, \pm 1, \ldots, \pm\infty, \qquad m^2 + n^2 \neq 0.$$

Using Result 4.1 and substituting $c_{0,0}^z = \pi c$ and $\lambda_{m,n} = -4(m^2 + n^2)c_2$ [Eq. (4.36)], the above equations can be simplified as follows:

$$z_{0,0}(t_{\text{dep}}) = z_{0,0}(t_K) + \pi c(t_K)t_c + \pi c(t_{K+1})(t_{\text{dep}} - t_{K+1})$$
$$+ \bar{\theta}_{0,0}(t_K) + \hat{\theta}_{0,0}(t_{K+1}),$$

$$z_{m,n}(t) = e^{-4(m^2 + n^2)\{c_2(t_K)t_c + c_2(t_{K+1})[t_{\text{dep}} - t_{K+1}]\}} z_{m,n}(t_K)$$
$$+ \bar{\theta}_{m,n}(t_K) + \hat{\theta}_{m,n}(t_{K+1}),$$

$$m, n = 0, \pm 1, \ldots, \pm\infty, \qquad m^2 + n^2 \neq 0,$$

$$\tag{5.19}$$

where $\bar{\theta}_{0,0}(t_K)$, $\hat{\theta}_{0,0}(t_{K+1})$, $\bar{\theta}_{m,n}(t_K)$, and $\hat{\theta}_{m,n}(t_{K+1})$ $(m^2 + n^2 \neq 0)$ are independent Gaussian random numbers with zero mean, and their covariances can be expressed as follows:

$$\langle \bar{\theta}_{0,0}(t_K)\bar{\theta}_{0,0}^*(t_K)\rangle = \varsigma^2(t_K)t_c,$$

$$\langle \hat{\theta}_{0,0}(t_{K+1})\hat{\theta}_{0,0}^*(t_{K+1})\rangle = \varsigma^2(t_{K+1})(t_{\text{dep}} - t_{K+1}),$$

$$\langle \bar{\theta}_{m,n}(t_K)\bar{\theta}_{m,n}^*(t_K)\rangle = e^{-8(m^2+n^2)c_2(t_{K+1})t_c}\varsigma^2(t_K)$$

$$\times \frac{e^{-8(m^2+n^2)c_2(t_K)t_c} - 1}{-8(m^2+n^2)c_2(t_K)},$$

$$\langle \hat{\theta}_{m,n}(t_{K+1})\hat{\theta}_{m,n}^*(t_{K+1})\rangle = \varsigma^2(t_{K+1})\frac{e^{-8(m^2+n^2)c_2(t_{K+1})(t_{dep}-t_{K+1})} - 1}{-8(m^2+n^2)c_2(t_{K+1})},$$

$$m, n = 0, \pm 1, \dots, \pm\infty, \quad m^2 + n^2 \neq 0. \tag{5.20}$$

Therefore, the quantities that directly relate to thickness and roughness estimation, $\langle z_{0,0}(t_{\text{dep}})\rangle$, $\langle z_{m,n}(t_{\text{dep}})\rangle$, and $\langle z_{m,n}(t_{\text{dep}})z_{m,n}^*(t_{\text{dep}})\rangle$, can be derived as follows:

$$\langle z_{0,0}(t_{\text{dep}})\rangle = z_{0,0}(t_K) + \pi c(t_K)t_c + \pi c(t_{K+1})(t_{\text{dep}} - t_{K+1}),$$

$$\langle z_{m,n}(t_{\text{dep}})\rangle = e^{-4(m^2+n^2)[c_2(t_K)t_c + c_2(t_{K+1})(t_{\text{dep}}-t_{K+1})]}z_{m,n}$$

$$\times (t_K),$$

$$\langle z_{m,n}(t_{\text{dep}})z_{m,n}^*(t_{\text{dep}})\rangle = \langle z_{m,n}(t_{\text{dep}})\rangle\langle z_{m,n}(t_{\text{dep}})\rangle^* + \langle \bar{\theta}_{m,n}(t_K)\bar{\theta}_{m,n}^*(t_K)\rangle$$

$$+ \langle \hat{\theta}_{m,n}(t_{K+1})\hat{\theta}_{m,n}^*(t_{K+1})\rangle,$$

$$m, n = 0, \pm 1, \dots, \pm\infty, \quad m^2 + n^2 \neq 0. \tag{5.21}$$

Accordingly, the expected final film thickness can be expressed as follows:

$$\langle \bar{h}_{\text{final}}(t_K)\rangle = \frac{\langle z_{0,0}(t_{\text{dep}})\rangle}{\pi} = \frac{z_{0,0}(t_K)}{\pi} + c(t_K)t_c + c(t_{K+1})(t_{\text{dep}} - t_{K+1}). \tag{5.22}$$

Also, by substituting Eq. (5.21) into Eq. (4.63), the expected final surface roughness can be derived as follows:

$$\langle r_{\text{final}}^2(t_K) \rangle = \frac{1}{\pi^2} \sum_{m,n=-\infty;\ m^2+n^2\neq0}^{\infty} \langle z_{m,n}(t_{\text{dep}}) z_{m,n}^*(t_{\text{dep}}) \rangle$$

$$= \frac{1}{\pi^2} \sum_{m,n=-\infty;\ m^2+n^2\neq0}^{\infty} [\langle z_{m,n}(t_{\text{dep}}) \rangle \langle z_{m,n}(t_{\text{dep}}) \rangle^*$$

$$+ \langle \bar{\theta}_{m,n}(t_K) \bar{\theta}_{m,n}^*(t_K) \rangle + \langle \hat{\theta}_{m,n}(t_{K+1}) \hat{\theta}_{m,n}^*(t_{K+1}) \rangle]$$

$$= \frac{1}{\pi^2} \sum_{m,n=-\infty;\ m^2+n^2\neq0}^{\infty} \Big\{ z_{m,n}(t_K) z_{m,n}^*(t_K)$$

$$\times\ e^{-8(m^2+n^2)[c_2(t_K)t_c + c_2(t_{K+1})(t_{\text{dep}}-t_{K+1})]}$$

$$+\ e^{-8(m^2+n^2)c_2(t_{K+1})t_c} \varsigma^2(t_K) \frac{e^{-8(m^2+n^2)c_2(t_K)t_c} - 1}{-8(m^2+n^2)c_2(t_K)}$$

$$+\ \varsigma^2(t_{K+1}) \frac{e^{-8(m^2+n^2)c_2(t_{K+1})(t_{\text{dep}}-t_{K+1})} - 1}{-8(m^2+n^2)c_2(t_{K+1})} \Big\}.$$

$$(5.23)$$

Since the computation of the above equation involves infinite summations, it cannot be calculated directly in practice. A finite-dimensional approximation, which only uses the first ($\pm N$th,$\pm N$th) states, is used for the computation and is of the following form:

$$\langle r_{\text{final}}^2(t_K) \rangle = \frac{1}{\pi^2} \sum_{m,n=-N;\ m^2+n^2\neq0}^{N} \Big\{ z_{m,n}(t_K) z_{m,n}^*(t_K)$$

$$\times\ e^{-8(m^2+n^2)[c_2(t_K)t_c + c_2(t_{K+1})(t_{\text{dep}}-t_{K+1})]}$$

$$+\ e^{-8(m^2+n^2)c_2(t_{K+1})t_c} \varsigma^2(t_K) \frac{e^{-8(m^2+n^2)c_2(t_K)t_c} - 1}{-8(m^2+n^2)c_2(t_K)}$$

$$+\ \varsigma^2(t_{K+1}) \frac{e^{-8(m^2+n^2)c_2(t_{K+1})(t_{\text{dep}}-t_{K+1})} - 1}{-8(m^2+n^2)c_2(t_{K+1})} \Big\}.$$

$$(5.24)$$

Here we note that this finite-dimensional approximation can be improved by utilizing an upper bound for the residue of the infinite summation derived following the method we proposed in our previous chapter (see Section 5.2.2 for a discussion on the convergence property of the infinite series and the determination of N for a desired approximation precision). However, such an improvement is not adopted in this chapter for simplicity. Moreover, instead of direct truncation of the system of infinite-dimensional stochastic ODEs, more advanced reduction techniques can be used, especially when the stochastic PDE model is nonlinear (see [12, 32] for results on nonlinear model reduction of parabolic PDEs). Therefore, the values of $T(t_{K+1})$ and $W(t_{K+1})$ are determined at each sampling time interval by solving, in the control time interval, the following optimization problem:

$$\min_{W(t_{K+1}),T(t_{K+1})} J(t_K) = q_h(h_{\text{set}} - \langle \bar{h}_{\text{final}}(t_K) \rangle)^2 + q_r(r_{\text{set}}^2 - \langle r_{\text{final}}^2(t_K) \rangle)^2 \tag{5.25}$$

subject to

$$\langle \bar{h}_{\text{final}}(t_K) \rangle = \frac{z_{0,0}(t_K)}{\pi} + c(t_K)t_c + c(t_{K+1})(t_{\text{dep}} - t_{K+1}), \tag{5.26}$$

$$\langle r_{\text{final}}^2(t_K) \rangle = \frac{1}{\pi^2} \sum_{m,n=-N;\, m^2+n^2 \neq 0}^{N} \left\{ z_{m,n}(t_K)z_{m,n}^*(t_K) \right.$$
$$\times\; e^{-8(m^2+n^2)[c_2(t_K)t_c + c_2(t_{K+1})(t_{\text{dep}} - t_{K+1})]}$$
$$+\; e^{-8(m^2+n^2)c_2(t_{K+1})t_c} \varsigma^2(K) \frac{e^{-8(m^2+n^2)c_2(t_K)t_c} - 1}{-8(m^2+n^2)c_2(t_K)}$$
$$+\; \varsigma^2(t_{K+1}) \left. \frac{e^{-8(m^2+n^2)c_2(t_{K+1})(t_{\text{dep}} - t_{K+1})} - 1}{-8(m^2+n^2)c_2(t_{K+1})} \right\}, \tag{5.27}$$

$$c(t_{K+1}) = W(t_{K+1}) \left[1 - \frac{k_w}{W(t_{K+1})}^{a_w} e^{-k_B T}(t_{K+1})/E_w \right], \tag{5.28}$$

$$c_2(t_{K+1}) = \frac{k_c}{k_{\max}^2 W(t_{K+1})^{a_c} e^{-k_B T(t_{K+1})/E_c}}, \tag{5.29}$$

$$\varsigma^2(t_{K+1}) = \frac{\pi^2}{k_{\max}^2} W(t_{K+1}) \left\{ 1 + \frac{e^{[a_t + k_t W(t_{K+1})]T(t_{K+1})}}{e^{a_v} + k_v W(t_{K+1})} \right\}, \tag{5.30}$$

$$c(t_{K+1}) \geq c_{\min}, \tag{5.31}$$

$$T_{z_{\min}} \leq T(t_{K+1}) \leq T_{\max}, \tag{5.32}$$

$$W_{\min} \leq W(t_{K+1}) \leq W_{\max}, \tag{5.33}$$

where q_h and q_r are the weights of the penalties on thickness and roughness, respectively, $c_{\min}$ is the minimum growth rate, $T_{\min}, T_{\max}, W_{\min}$ and $W_{\max}$ are the lowest and highest substrate temperature, and the lowest and highest adsorption rate, respectively. In this study, we use $q_h = 1/h_{\text{set}}^2$, $q_r = 1/r_{\text{set}}^2$, $c_{\min} = 0.1h_{\text{set}}/t_{\text{dep}}$, $T_{\min} = 400$ K, $T_{\max} = 900$ K, $W_{\min} = 0.11/\text{s}$ and $W_{\max} = 2.01/\text{s}$.

We note that J corresponds to the difference between the square of the desired final surface roughness r_{set} and the square of the estimated final surface roughness

$\langle r^2_{\text{final}} \rangle$ computed based on the current states $z_{m,n}(t_K)$. We choose to minimize the difference of the squares of the surface roughness, i.e., the mean square of the surface height, to simplify the calculation. The optimization problem is solved using a standard sequential quadratic programming (SQP) method described in [45]. Also, since Eq. (5.27) is a finite-dimensional approximation of the predicted final surface roughness, to achieve a control precision ϵ, m should be chosen large enough for each optimization computation so that the approximation error is less than ϵ.

Closed-Loop Simulations

A kMC simulation using a lattice size of 100×100 is used to simulate the thin-film deposition process and t_c is set to be 1 s. The dimension of the finite-dimensional approximation of the stochastic PDE used for optimization is $N = 10$.

Figure 5.6 shows the surface roughness and substrate temperature profiles of a closed-loop deposition process with initial substrate temperature $T = 610$ K and adsorption rate $W = 1.0$ s^{-1} (these initial values are picked such that, with process parameters fixed at these levels throughout the deposition, the final thickness and surface roughness of the deposited film are quite different from the desired values). Figure 5.7 shows the thin-film thickness and surface adsorption rate profiles of this closed-loop deposition. The control objective is to control the thin-film thickness to 100 ML (monolayers) and to drive the final surface roughness to 1.5 ML at the

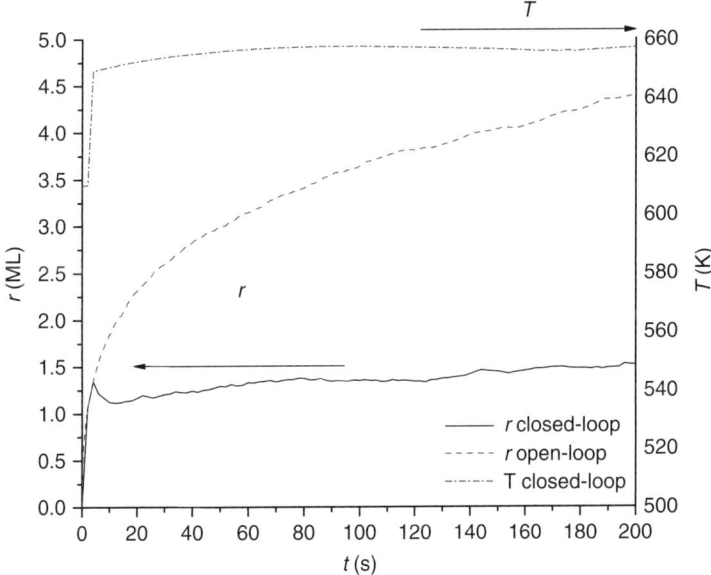

Fig. 5.6. Surface roughness and substrate temperature profiles of a 200-s closed-loop deposition process with a thickness set point of 100 ML and a final roughness set point $r_{\text{set}} = 1.5$ ML; the initial deposition conditions are $T = 610$ K and $W = 1.0$ ML/s.

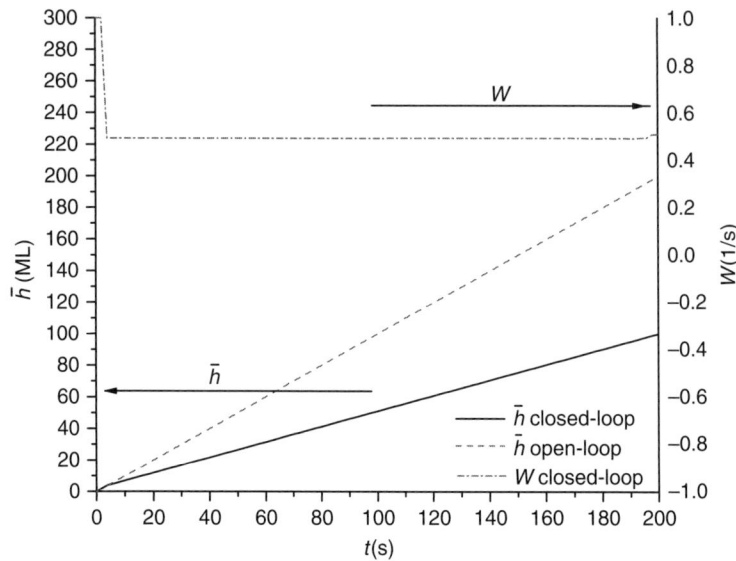

Fig. 5.7. Thickness and surface adsorption rate profiles of a 200-s closed-loop deposition process with a thickness setpoint of 100 ML and a final roughness set point $r_{set} = 1.5$ ML; the initial deposition conditions are $T = 610$ K and $W = 1.0$ ML/s.

end of the 200-s deposition. It can be seen that both the film thickness and the final surface roughness are controlled at the desired levels simultaneously while an open-loop deposition with the same initial deposition condition would lead to a 100% higher film thickness and a 100% higher final surface roughness as shown in Figs. 5.6 and 5.7.

Figure 5.8 shows the final surface roughness histogram of the thin films deposited using 100 different closed-loop depositions targeting a thin-film thickness of 100 ML and final surface roughness of 1.65 ML and 100 different open-loop depositions with fixed substrate temperature and surface adsorption rate. The average roughness of the thin films deposited by open-loop depositions is 1.52 ML, which is quite close to the average roughness of the thin films deposited by the closed-loop depositions (1.64 ML). However, the variance among the thin films from different open-loop deposition runs is over 300% higher than that of closed-loop depositions even though no process disturbance is considered in the simulations.

Such a large variance among the films deposited by open-loop deposition can be attributed to the stochastic nature of the thin-film growth process itself. Although optimal profiles of adsorption rate and substrate temperature, i.e., a well-prescribed process recipe, can be determined for the open-loop deposition, so that the average final thickness and surface roughness of the deposited films are very close to the desired levels, the stochasticity of the film growth cannot be effectively handled by the predetermined process recipes (implemented in an open-loop fashion) and, therefore, results in significant film roughness variance. On the other hand, in closed-loop depositions, as is demonstrated in the simulation, feedback control is able to effectively

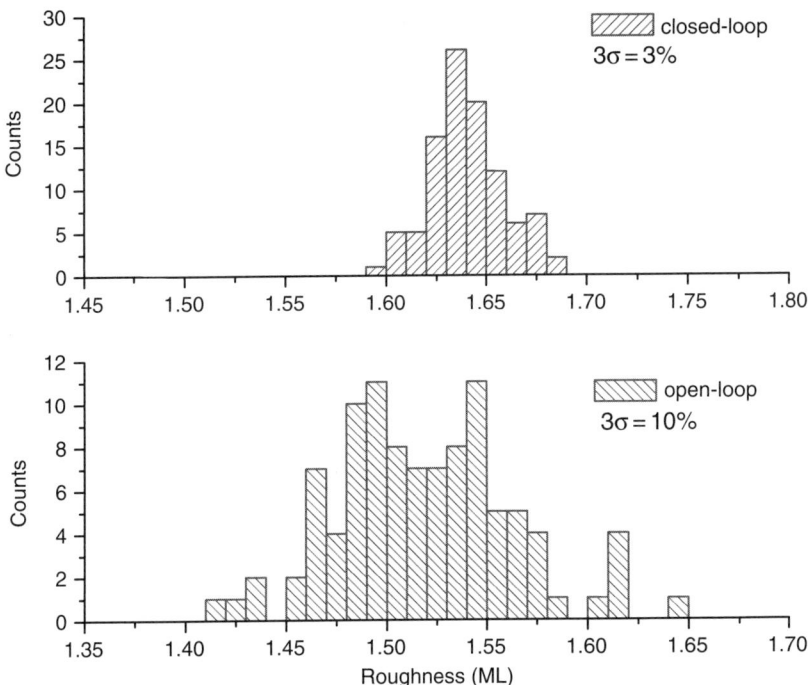

Fig. 5.8. Histogram of final surface roughness of 100 closed-loop and 100 open-loop thin-film depositions.

compensate for the stochasticity of the process and, therefore, significantly reduce the film variance and outperform the recipe-based, open-loop deposition.

Remark 5.1. Since the control action is computed using closed-form equations in the MPC design, the computational cost is proportional to the number of states used, but independent of the optimization horizon $t_{dep} - t$. Nevertheless, even for a lattice size that corresponds to the largest physical dimension of the sampling area that can be achieved by common surface measurement techniques (i.e., a few microns), such computation can still be completed within the control interval using currently available computing power. On the other hand, such a task is almost impossible to achieve using a kMC code, whose computational cost is on the order of $k_{max}^4(t_{dep} - t)$ for merely a single run. Furthermore, we note that the evaluation of each state is independent of other states and, therefore, can be executed in parallel, while the kMC code, being a serial calculation, is unsuitable for parallel processing.

5.3 Linear Covariance Control Using Stochastic PDEs

When spatially distributed sensing and actuation are available, the surface roughness can be regulated using the covariance control technique. In this section, we present a method for covariance control of the surface roughness based on linear stochastic PDEs. The following second-order linear stochastic PDE is used to present our method:

$$\frac{\partial h}{\partial t} = \nu \frac{\partial^2 h}{\partial x^2} + \xi(x, t), \tag{5.34}$$

where $\nu > 0$ is a constant, $x \in [-\pi, \pi]$ is the spatial coordinate, t is the time, $h(x,t)$ is the height of the surface at position x and time t, and $\xi(x,t)$ is a Gaussian noise with zero mean and covariance

$$\langle \xi(x, t)\xi(x', t') \rangle = \varsigma^2 \delta(x - x')\delta(t - t'), \tag{5.35}$$

where ς^2 is a constant, $\delta(\cdot)$ is the Dirac function, and $\langle \cdot \rangle$ denotes the expected value. Note that the noise covariance depends on both space x and time t.

The surface roughness, r, is represented by the standard deviation of the surface from its average height. The definition is similar to Eq. (3.2) and is given as follows:

$$r(t) = \sqrt{\frac{1}{2\pi} \int_{-\pi}^{\pi} [h(x, t) - \bar{h}(t)]^2 dx}, \tag{5.36}$$

where $\bar{h}(t) = (1/2\pi) \int_{-\pi}^{\pi} h(x, t)dx$ is the average surface height.

Our objective is to control the surface roughness of a simple thin-film growth process described by Fig. (5.9). The controller design is based on the stochastic partial differential equation (SPDE) model of the process [Eqs. (5.34) and (5.35)]. To do this, we formulate a distributed control problem in the spatial domain $[-\pi, \pi]$. The control problem is described by the following SPDE:

$$\frac{\partial h}{\partial t} = \nu \frac{\partial^2 h}{\partial x^2} + \sum_{i=1}^{p} b_i(x)u_i(t) + \xi(x, t) \tag{5.37}$$

subject to periodic boundary conditions

$$\frac{\partial^j h}{\partial x^j}(-\pi, t) = \frac{\partial^j h}{\partial x^j}(\pi, t), \quad j = 0, 1, \tag{5.38}$$

and with the initial condition

$$h(x, 0) = h_0(x), \tag{5.39}$$

where u_i is the ith manipulated input, p is the number of manipulated inputs, and b_i is the ith actuator distribution function [i.e., b_i determines how the control action computed by the ith control actuator, u_i, is distributed (e.g., point or distributed actuation) in the spatial interval $(-\pi, \pi)$].

To study the dynamics of Eq. (5.37), we initially consider the eigenvalue problem of the linear operator in Eq. (5.37), which takes the form

$$A\bar{\phi}_n(x) = \nu \frac{d^2\bar{\phi}_n(x)}{dx^2} = \lambda_n \bar{\phi}_n(x), \quad n = 1, \ldots, \infty,$$

$$\frac{d^j \bar{\phi}_n}{dx^j}(-\pi) = \frac{d^j \bar{\phi}_n}{dx^j}(\pi), \quad j = 0, 1,$$

(5.40)

where λ_n denotes an eigenvalue and $\bar{\phi}_n$ denotes an eigenfunction. A direct computation of the solution of the above eigenvalue problem yields $\lambda_0 = 0$ with $\psi_0 = 1/\sqrt{2\pi}$, and $\lambda_n = -\nu n^2$ (λ_n is an eigenvalue of multiplicity two) with eigenfunctions $\phi_n = (1/\sqrt{\pi})\sin(nx)$ and $\psi_n = (1/\sqrt{\pi})\cos(nx)$ for $n = 1, \ldots, \infty$. From the solution of the eigenvalue problem shown in Eq. (5.40), it follows that for a fixed value of $\nu > 0$, the distance between two consecutive eigenvalues (i.e., λ_n and λ_{n+1}) increases as n increases. Furthermore, the eigenspectrum of the operator A in Eq. (5.40), $\sigma(A)$, can be partitioned as $\sigma(A) = \sigma_1(A) \bigcup \sigma_2(A)$, where $\sigma_1(A)$ contains the first m (with m finite) eigenvalues [i.e., $\sigma_1(A) = \{\lambda_1, \ldots, \lambda_m\}$] and $\sigma_2(A)$ contains the remaining infinite number of eigenvalues [i.e., $\sigma_2(A) = \{\lambda_{m+1}, \ldots\}$].

To present the method that we use to control the stochastic PDE in Eq. (5.37), we first derive stochastic ODE approximations in Eq. (5.37) using modal decomposition. To this end, we first expand the solution in Eq. (5.37) in an infinite series in terms of the eigenfunctions of the operator in Eq. (5.40) as follows:

$$h(x, t) = \sum_{n=1}^{\infty} \alpha_n(t)\phi_n(x) + \sum_{n=0}^{\infty} \beta_n(t)\psi_n(x),$$

(5.41)

where $\alpha_n(t)$ and $\beta_n(t)$ are time-varying coefficients. Substituting the above expansion for the solution, $h(x, t)$, into Eq. (5.37) and taking the inner product with the adjoint eigenfunctions, $\phi_n^*(z) = (1/\sqrt{\pi})\sin(nz)$ and $\psi_n^*(z) = (1/\sqrt{\pi})\cos(nz)$, the following system of infinite stochastic ODEs is obtained:

$$\frac{d\alpha_n}{dt} = -\nu n^2 \alpha_n + \sum_{i=1}^{p} b_{i\alpha_n} u_i(t) + \xi_\alpha^n(t),$$

$$\frac{d\beta_n}{dt} = -\nu n^2 \beta_n + \sum_{i=1}^{p} b_{i\beta_n} u_i(t) + \xi_\beta^n(t), \quad n = 1, \ldots, \infty,$$

(5.42)

where $b_{i\alpha_n} = \int_{-\pi}^{\pi} \phi_n^*(x)b_i(x)dx$, $b_{i\beta_n} = \int_{-\pi}^{\pi} \psi_n^*(x)b_i(x)dx$,

$\xi_\alpha^n(t) = \int_{-\pi}^{\pi} \xi(x, t)\phi_n^*(x)dx$, $\xi_\beta^n(t) = \int_{-\pi}^{\pi} \xi(x, t)\psi_n^*(x)dx$.

The covariances of $\xi_\alpha^n(t)$ and $\xi_\beta^n(t)$ can be computed using Result 4.1 as $\langle \xi_\alpha^n(t)\xi_\alpha^n(t')\rangle = \varsigma^2\delta(t - t')$ and $\langle \xi_\beta^n(t)\xi_\beta^n(t')\rangle = \varsigma^2\delta(t - t')$.

In this work, the controlled variable is the expected value of surface roughness, $\langle r^2 \rangle$. According to Eq. (5.41), we have $\bar{h}(t) = \beta_0(t)\psi_0$. Therefore, $\langle r^2 \rangle$ can be rewritten in terms of α_n and β_n as follows:

$$\langle r^2 \rangle = \frac{1}{2\pi} \left\langle \int_{-\pi}^{\pi} (h(x,t) - \bar{h}(t))^2 dx \right\rangle$$

$$= \frac{1}{2\pi} \left\langle \int_{-\pi}^{\pi} \sum_{i=1}^{\infty} \left[\alpha_i^2(t)\phi_i^2(x) + \beta_i^2(t)\psi_i^2(x) \right] dx \right\rangle \tag{5.43}$$

$$= \frac{1}{2\pi} \left\langle \sum_{i=1}^{\infty} (\alpha_i^2 + \beta_i^2) \right\rangle = \frac{1}{2\pi} \sum_{i=1}^{\infty} \left[\langle \alpha_i^2 \rangle + \langle \beta_i^2 \rangle \right].$$

Therefore, the surface roughness control problem for the stochastic PDE system in Eq. (5.37) is formulated as that of controlling the covariance of the states α_n and β_n of the system of infinite stochastic ODEs in Eq. (5.42).

5.3.1 Model Reduction

Due to its infinite-dimensional nature, the system in Eq. (5.42) cannot be directly used for the design of controllers that can be implemented in practice (i.e., the practical implementation of controllers that are designed on the basis of this system will require the computation of infinite sums, which cannot be done by a computer). Instead, we base the controller design on finite-dimensional approximations of this system. Subsequently, we will show that the resulting controller will enforce the desired control objective in the closed-loop infinite-dimensional system.

Specifically, we rewrite the system in Eq. (5.42) as follows:

$$\frac{dx_s}{dt} = \Lambda_s x_s + B_s u + \xi_s,$$

$$\frac{dx_f}{dt} = \Lambda_f x_f + B_f u + \xi_f, \tag{5.44}$$

where $x_s = [\alpha_1 \cdots \alpha_m \ \beta_1 \cdots \beta_m]^T$, $x_f = [\alpha_{m+1}\beta_{m+1}\cdots]^T$, $\Lambda_s = \text{diag}[-\nu \cdots - m^2\nu - \nu \cdots - m^2\nu]$, $\Lambda_f = \text{diag}[-(m+1)^2\nu - (m+1)^2\nu \cdots]$, $u = [u_1 \cdots u_p]^T$, $\xi_s = [\xi_\alpha^1 \cdots \xi_\alpha^m \ \xi_\beta^1 \cdots \xi_\beta^m]^T$, and $\xi_f = [\xi_\alpha^{m+1}\xi_\beta^{m+1}\cdots]^T$

$$B_s = \begin{bmatrix} b_{1\alpha_1} & \cdots & b_{p\alpha_1} \\ \vdots & \ddots & \vdots \\ b_{1\alpha_m} & \cdots & b_{p\alpha_m} \\ b_{1\beta_1} & \cdots & b_{p\beta_1} \\ \vdots & \ddots & \vdots \\ b_{1\beta_m} & \cdots & b_{p\beta_m} \end{bmatrix}, \quad B_f = \begin{bmatrix} b_{1\alpha_{m+1}} & \cdots & b_{p\alpha_{m+1}} \\ b_{1\beta_{m+1}} & \cdots & b_{p\beta_{m+1}} \\ b_{1\alpha_{m+2}} & \cdots & b_{p\alpha_{m+2}} \\ b_{1\beta_{m+2}} & \cdots & b_{p\beta_{m+2}} \\ \vdots & \vdots & \vdots \end{bmatrix}. \tag{5.45}$$

We note that the subsystem x_f in Eq. (5.44) is infinite-dimensional.

Neglecting the x_f subsystem, the following $2m$-dimensional system is obtained:

$$\frac{d\tilde{x}_s}{dt} = \Lambda_s \tilde{x}_s + B_s u + \xi_s, \tag{5.46}$$

where the tilde symbol in $\tilde{x}_s$ denotes that this state variable is associated with a finite-dimensional system.

5.3.2 Feedback Control Design

We design the state feedback controller on the basis of the finite-dimensional system in Eq. (5.46). To simplify our development, we assume that $p = 2m$ and we pick the actuator distribution functions such that B_s^{-1} exists. The state feedback control law then takes the form

$$u = B_s^{-1} \left(\Lambda_{cs} - \Lambda_s \right) \tilde{x}_s, \tag{5.47}$$

where the matrix Λ_{cs} contains the desired poles of the closed-loop system; $\Lambda_{cs} = \mathrm{diag}[\lambda_{c\alpha 1} \cdots \lambda_{c\alpha m} \lambda_{c\beta 1} \cdots \lambda_{c\beta m}]$, and $\lambda_{c\alpha i}$ and $\lambda_{c\beta i}$ ($1 \leq i \leq m$) are desired poles of the closed-loop, finite-dimensional system, which can be computed from the desired closed-loop surface roughness level.

We first analyze the dependence of the covariances of the states α_n and β_n ($n = 1, \ldots, m$) on the poles of the finite-dimensional system in Eq. (5.46). Then, we will show in the next subsection that the surface roughness of the infinite-dimensional system in Eq. (5.42) can be controlled to a desired level by using the state feedback controller in Eq.(5.47), which uses only a finite number of actuators.

By applying the controller in Eq. (5.47) to the system in Eq. (5.46), the closed-loop system takes the form

$$\frac{d\tilde{x}_s}{dt} = \Lambda_{cs} \tilde{x}_s + \xi_s(t). \tag{5.48}$$

To analyze the effect of the feedback controller on the covariance of the state $\tilde{x}_s$, we discretize Eq. (5.48) in the time domain, using Δt as the time step, as follows:

$$X_s(k+1) = G_{cs} X_s(k) + \zeta_s(k), \quad k = 0, \ldots, \infty, \tag{5.49}$$

where $X_s(k) = \tilde{x}_s(k\Delta t)$, and $G_{cs} = e^{\Lambda_{cs} \Delta t}$, $\zeta_s(k) = \int_{k\Delta t}^{(k+1)\Delta t} e^{\Lambda_{cs}((k+1)\Delta t - t)} \xi_s(t)dt$. According to [1, Chapter 3], if all eigenvalues of G_{cs} are within the unit circle on the complex plane, the covariance matrix of $X_s(k)$, $P(k) = \langle X_s(k)X_s(k)^T \rangle$, converges to $P(\infty)$, which is the solution to the following equation:

$$P(\infty) = G_{cs}P(\infty)G_{cs}^T + R_1, \tag{5.50}$$

where $R_1 = \langle \zeta_s \zeta_s^T \rangle$. Eq. (5.50) cannot be solved, in general, analytically. However, for the specific deposition system considered in this work, the analytical solution for $P(\infty)$ can be obtained as follows:

$$P(\infty) = \begin{bmatrix} P_\alpha(\infty) & 0 \\ 0 & P_\beta(\infty) \end{bmatrix}, \tag{5.51}$$

where $P_\alpha(\infty) = \text{diag}[\langle \alpha_1^2(\infty) \rangle \cdots \langle \alpha_m^2(\infty) \rangle]$, and $P_\beta(\infty) = \text{diag}[\langle \beta_1^2(\infty) \rangle \cdots \langle \beta_m^2(\infty) \rangle]$. Using Result 5.1, $\langle \alpha_n^2(\infty) \rangle$ and $\langle \beta_n^2(\infty) \rangle$ $(n = 1, \ldots, m)$ can be computed by using the following expressions:

$$\langle \alpha_n^2(\infty) \rangle = -\frac{\varsigma^2}{2\lambda_{c\alpha_n}}, \quad \langle \beta_n^2(\infty) \rangle = -\frac{\varsigma^2}{2\lambda_{c\beta_n}}. \tag{5.52}$$

From Eq. (5.52), we can see that by assigning the closed-loop poles $\lambda_{c\alpha_n}$ and $\lambda_{c\beta_n}$ $(n = 1, \ldots, m)$ at desired locations, the covariances of the states α_n and β_n $(n = 1, \ldots, m)$ can be controlled to desired levels. Therefore, according to Eq. (5.43), the contribution to the surface roughness from the finite-dimensional system in Eq. (5.46) can be controlled to the desired level.

5.3.3 Analysis of the Closed-Loop Infinite-Dimensional System

In this subsection, we show that when the state feedback controller in Eq. (5.47) is used to manipulate the poles of the finite-dimensional system in Eq. (5.46), the contribution to the surface roughness from the α_f and β_f subsystem of the system in Eq. (5.44) is bounded and can be made arbitrarily small by increasing the dimension of the x_s subsystem.

By applying the feedback controller in Eq. (5.47) into the infinite-dimensional system in Eq. (5.44), we obtain the following closed-loop system:

$$\frac{dx_s}{dt} = \Lambda_{cs}x_s + \xi_s,$$

$$\frac{dx_f}{dt} = \Lambda_\epsilon x_s + \Lambda_f x_f + \xi_f, \tag{5.53}$$

where $\Lambda_\epsilon = B_f B_s^{-1}(\Lambda_{cs} - \Lambda_s)$.

The boundedness of the state of the above system follows directly from the stability of the matrices Λ_{cs} and Λ_f and the structure of the system, where the x_s subsystem is independent of the x_f state (see [32, 28] for results and techniques for analyzing the stability properties of such systems).

Due to the structure of the eigenspectrum of operator A, the effect of the control action computed from Eq. (5.47) to the poles of the x_f subsystem can be reduced by increasing m. Therefore, by picking m sufficiently large, the $\Lambda_\epsilon x_s$ term can be made very small compared to the $\Lambda_f x_f$ term, thus, the closed-loop system in Eq. (5.53) can be adequately described by the following system:

$$\frac{dx_s}{dt} = \Lambda_{cs}x_s + \xi_s,$$

$$\frac{dx_f}{dt} = \Lambda_f x_f + \xi_f. \tag{5.54}$$

On the basis of the above system, it can be shown that the covariance of the state of the x_f subsystem converges to $[\langle \alpha_{m+1}^2(\infty) \rangle \langle \beta_{m+1}^2(\infty) \rangle \cdots \cdots]$, where

$$\langle \alpha_n^2(\infty) \rangle = \frac{\varsigma^2}{2n^2\nu}, \quad \langle \beta_n^2(\infty) \rangle = \frac{\varsigma^2}{2n^2\nu}, \quad n > m. \tag{5.55}$$

Therefore, for m sufficiently large, the overall contribution to the surface roughness from the x_f subsystem in Eq. (5.44) can be computed as follows:

$$\frac{\varsigma}{\sqrt{2\pi(m+1)\nu}} < \sqrt{\frac{1}{2\pi} \sum_{n=m+1}^{\infty} \left[\frac{\varsigma^2}{\nu n^2} \right]} < \frac{\varsigma}{\sqrt{2\pi m\nu}}. \tag{5.56}$$

Clearly, as $m \to \infty$, the contribution to the surface roughness from the α_f and β_f subsystem goes to zero.

In summary, under the controller in Eq. (5.47), the closed-loop surface roughness, for m sufficiently large, can be adequately described by the following expression:

$$\langle r^2 \rangle = \frac{\varsigma^2}{2\pi} \left(\lambda^* + \sum_{n=m+1}^{\infty} \frac{1}{\nu n^2} \right), \tag{5.57}$$

where $\lambda^* = \sum_{i=1}^{m} [(-1/2\lambda_{c\alpha_i}) - (1/2\lambda_{c\beta_i})]$.

Remark 5.2. Note that in order to regulate the surface roughness to a desired level, $\langle r_d^2 \rangle$, the number of actuators should be large enough so that the value of $\langle r_d^2 \rangle$ is achievable. Specifically, the number of actuators m should be selected such that the following inequality holds:

$$\langle r_d^2 \rangle > \frac{\varsigma^2}{2\pi} \left(\sum_{n=m+1}^{\infty} \frac{1}{\nu n^2} \right). \tag{5.58}$$

This is because the closed-loop stability requires that $\lambda_{c\alpha_i} < 0$ and $\lambda_{c\beta_i} < 0$ (for $i = 1, \ldots, m$), and, thus, $\lambda^* > 0$ in Eq. (5.57).

Remark 5.3. Note that to control the closed-loop surface roughness to $\langle r_d^2 \rangle$, we need to design a controller to assign the poles of the finite-dimensional system in Eq. (5.48) to appropriate values so that the following equation holds:

$$\lambda^* = \frac{2\pi(\langle r_d^2 \rangle - \langle r_f^2 \rangle)}{\varsigma^2}. \tag{5.59}$$

The controller that assigns the poles of the system in Eq. (5.48) to satisfy Eq. (5.59) is not unique. Consequently, for a fixed number of actuators, p, the controller that can regulate the closed-loop surface roughness to a desired level is also not unique. Furthermore, we note that robust control methods [27, 31], which utilize bounds of the noise terms, can be employed to design a controller that can achieve an arbitrary degree of attenuation of the effect of noise on the PDE system state.

Remark 5.4. Note that the expected value of the open-loop surface roughness converges to its steady-state value, $\langle r_{ol}^2 \rangle$, which can be computed as follows:

$$\langle r_{ol}^2 \rangle = \frac{\varsigma^2}{2\pi\nu} \left(\sum_{k=1}^{\infty} \frac{1}{k^2} \right). \tag{5.60}$$

If the closed-loop poles of the finite-dimensional system in Eq. (5.48) are written as $\lambda_{c\alpha_k} = -\nu n_{\alpha_k}^2$ and $\lambda_{c\beta_k} = -\nu n_{\beta_k}^2$ for $k = 1, \ldots, m$, the ratio of the expected value of the steady-state closed-loop surface roughness, $\langle r_{cl}^2 \rangle$, to that of the steady-state open-loop surface roughness, $\langle r_{ol}^2 \rangle$, can be computed as follows:

$$\frac{\langle r_{cl}^2 \rangle}{\langle r_{ol}^2 \rangle} = \frac{\displaystyle\sum_{k=1}^{m} \left(\frac{1}{n_{\alpha_k}^2} + \frac{1}{n_{\beta_k}^2} \right) + \sum_{k=m+1}^{\infty} \frac{1}{k^2}}{\displaystyle\sum_{k=1}^{\infty} \frac{1}{k^2}}. \tag{5.61}$$

This ratio is independent of the lattice size of the deposition system for the specific thin-film growth process under consideration. Therefore, if the control objective is to achieve a certain percentage of reduction of the value of surface roughness from that under open-loop operation, the number of actuators needed is independent of the lattice size of the deposition process.

5.3.4 Application to a Thin-Film Growth Process

Process Description

We consider a deposition process on a 1-dimensional lattice. In this process, particles land on the surface at rate r_a. The rules for the deposition are as follows: A site, l, is first randomly picked among the sites of the whole lattice and the deposition site is determined according to the following rules: (1) If the height of this site is lower than or equal to that of both nearest neighbors, this site is picked as the deposition site; (2) if the height of only one of the two nearest-neighbor sites is lower than that of the original site, deposition is on that site; (3) if the height of each of the nearest-neighbor sites is lower than that of the original site, the deposition site is randomly picked with equal probability between the two nearest-neighbor sites. A schematic of the rules of the deposition is shown in Fig. 5.9. There is no particle migration and desorption taking place on this process (see Chapter 3 and Sections 5.2.3 and 5.2.4 for film growth processes that involve these phenomena).

   In this section, we present an application of the proposed state feedback controller to the deposition process described in Fig. 5.9 to regulate the surface roughness to a desired level. Specifically, the deposition occurs on a lattice containing 1,000 sites. Therefore, $a = 0.00628$. The open-loop deposition rate for each site is $\bar{r}_a = 1\mathrm{s}^{-1}$. A 1000th -order stochastic ordinary differential equation approximation of the system in Eq. (5.37) is used to simulate the process (the use of higher-order

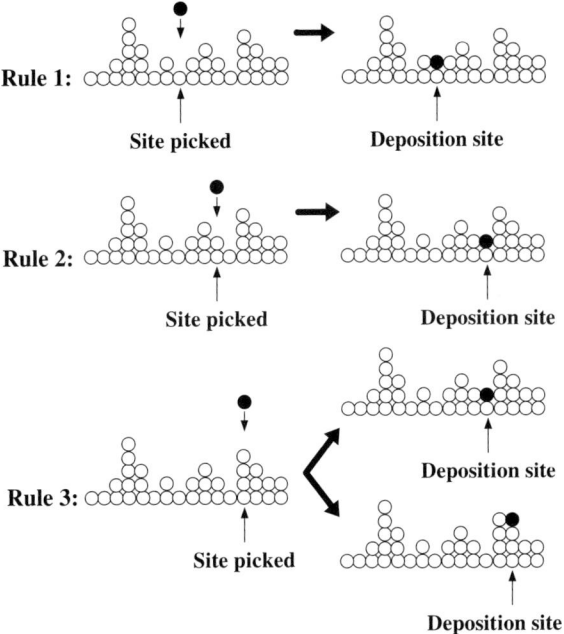

Fig. 5.9. Schematic of the rules of the deposition.

approximations led to identical numerical results, thereby implying that the following simulation runs are independent of the discretization). The δ function involved in the covariances of ξ_α^n and ξ_β^n is approximated by $1/\delta t$.

Open-Loop Dynamics

In the first simulation, we compare the expected value of the open-loop surface roughness of the deposition process from the solution of the stochastic PDE model in Eq. (5.34) to that from a kinetic Monte Carlo simulation. We use the kinetic Monte Carlo algorithm developed in [57] to simulate the process. First, a random number is generated to pick a site among all the sites on the 1D lattice. If the height of this site is lower than or equal to that of both nearest neighbors, this site is picked as the deposition site and the height of this site increases by a; if the height of only one of the two nearest-neighbor sites is lower than that of the original site, deposition is on that site and the height of that site increases by a; if the height of each of the nearest-neighbor sites is lower than that of the original site, a second random number is generated to randomly pick one of the two nearest neighbors with equal probability and the height of the picked site increases by a. Upon an executed event, a time increment, dt, is computed by $dt = (-\ln\zeta)/(N \times r_a)$, where ζ is a random number in the $(0, 1)$ interval and N is the total number of sites on the lattice.

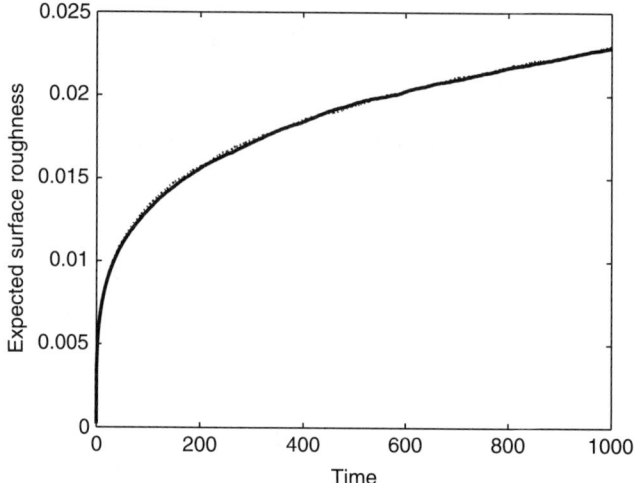

Fig. 5.10. Comparison of the open-loop profile of the expected surface roughness from the kinetic Monte Carlo simulator (solid line) and that from the solution of the SPDE, with adjusted covariance, using 1,000 modes (dotted line).

The profiles of expected surface roughness are obtained by averaging surface roughness profiles (either from the stochastic PDE or the kinetic Monte Carlo simulations) from 1,000 independent simulation runs using the same simulation parameters. According to [157], the model parameters ν and ς^2 are initially determined as $\nu = a^2 r_a = 3.944 \times 10^{-5}$ and $\varsigma^2 = a^3 r_a = 2.477 \times 10^{-7}$. Our simulation results show that if ς^2 is adjusted from 2.477×10^{-7} to 2.94×10^{-7}, the expected surface roughness profile computed from the stochastic PDE model better matches that from the kMC simulator (see [102] for a discussion on the adjustment of the model parameter). Figure 5.10 shows the simulation result. We can see that with this adjustment of ς, the two profiles are almost identical. Therefore, by slightly adjusting the covariance of the stochastic PDE model in Eq. (5.34), the model can adequately capture the evolution of surface roughness of the deposition process described in Fig. 5.9 obtained by the kinetic Monte Carlo simulations. Therefore, our control design will be based on the stochastic PDE model of the process with adjusted ς.

Closed-Loop Simulation

Subsequently, we design a state feedback covariance controller based on a 40th-order stochastic ODE approximation constructed by using the first 40 eigenmodes of the system in Eq. (5.42) with adjusted value of the covariance. Forty control actuators are used to control the system. The ith actuator distribution function is taken to be,

$$b_i(z) = \begin{cases} \dfrac{1}{\sqrt{\pi}}\sin(iz), & i = 1,\ldots,20, \\ \dfrac{1}{\sqrt{\pi}}\cos[(i-20)z], & i = 21,\ldots,40. \end{cases} \tag{5.62}$$

The expected open-loop surface roughness converges to 0.045, which can be computed using Eq. (5.60). The desired closed-loop surface roughness is 0.01 in this simulation, which is a 78% reduction compared to the open-loop surface roughness. Using Eq. (5.57), we design the state feedback controller such that $\lambda_{c\alpha_i} = \lambda_{c\beta_i} = -0.023$, for $i = 1, \ldots, 20$. Then, we apply the designed controller to the kinetic Monte Carlo model of the deposition process to control the surface roughness to the desired level.

The control action, u_i, can be implemented by manipulating the gas composition across the surface in a deposition process. Spatially controllable CVD reactors have been developed to enable across-wafer spatial control of surface gas composition during deposition [26]. In such a control problem formulation, the rate that particles land on the surface is spatially distributed and is computed by the controller. However, the value of r_a, which is used to calculate the values of ν and the covariance, ς, in the system in Eq. (5.37), corresponds to the adsorption rate under open-loop operation and is thus a constant. The contribution of the spatially distributed adsorption rate to the fluctuations of the surface height profile (e.g., the surface roughness) is captured by the term $\sum_{i=1}^{p} b_i(x)u_i(t)$. This control problem formulation is further supported by our simulation results that the controller designed based on the stochastic PDE model of the deposition process can be applied to the kinetic Monte Carlo model of the same deposition process to control the surface roughness to desired levels.

Specifically, the adsorption rate on site i at time t is determined according to the following expression:

$$r_a(i, t) = \bar{r}_a + \left(\sum_{j=1}^{40} b_j(z_i)u_j(t) \right) / a. \tag{5.63}$$

The following simulation algorithm is used to run the kinetic Monte Carlo simulations for the closed-loop system. First, a random number is generated to pick a site among all the sites on the 1D lattice; the probability that a surface site is picked is proportional to the adsorption rate on this site, which is computed using Eq. (5.63). If the height of this site is lower than or equal to that of both nearest neighbors, this site is picked as the deposition site and the height of this site increases by a; if the height of only one of the two nearest-neighbor sites is lower than that of the original site, deposition is on that site and the height of that site increases by a; if the height of each of the nearest-neighbor sites is lower than that of the original site, a second random number is generated to randomly pick one of the two nearest neighbors with equal probability and the height of the picked site increases by a. Upon an executed event, a time increment, dt, is computed by $dt = -\ln\zeta / \left(\sum_{i=1}^{N} r_a(i) \right)$, where ζ is a random number in the $(0, 1)$ interval and N is the total number of sites on the lattice. Once a particle is deposited, the first 40 states ($\alpha_1, \ldots, \alpha_{20}$ and $\beta_1, \ldots, \beta_{20}$) are updated and new control actions are computed to update the spatially distributed adsorption rate across the surface.

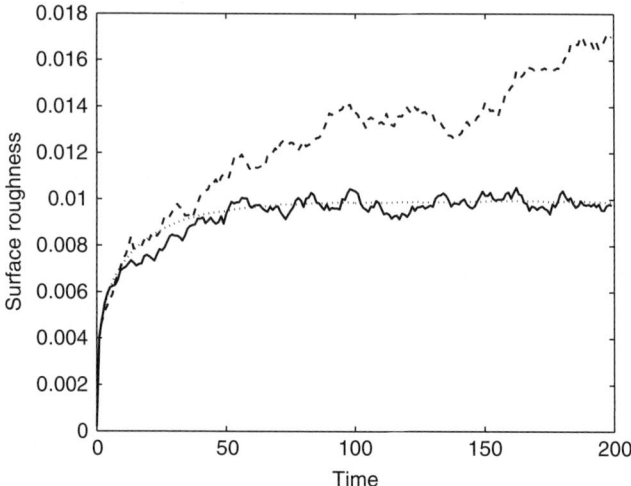

Fig. 5.11. Closed-loop simulation results by applying the controller designed based on the first 40 modes of the SPDE model to the kinetic Monte Carlo model. (a) The closed-loop surface roughness profile from one simulation run (solid line); (b) the expected closed-loop surface roughness profile (dotted line); and (c) the open-loop surface roughness profile from one simulation run (dashed line).

The closed-loop system simulation results are shown in Fig. 5.11. The dotted line shows the expected surface roughness, which is the average surface roughness profile obtained from 200 independent runs, under feedback control. We can see that the controller successfully drives the expected surface roughness to the desired level. The solid line shows the surface roughness profile under feedback control from one simulation run; due to the stochastic nature of the deposition process, stochastic fluctuations can be observed in the closed-loop surface roughness profile, but the surface roughness is very close to the set-point value under feedback control. For the sake of comparison, the dashed line shows a surface roughness profile from one open-loop simulation run. We can see that under feedback control, a much lower surface roughness can be achieved. Finally, we note that the proposed approach for controller design can be, in principle, applied to larger-scale deposition processes to control surface roughness. In such a case, the stochastic PDE model can be constructed by initially deriving a stochastic PDE model based on the transition rules and then fitting the model parameters based on the experimental roughness data from the specific deposition process.

5.4 Nonlinear Covariance Control Using Stochastic PDEs

Nonlinearities exist in many material preparation processes in which the surface evolution can be modeled by stochastic PDEs. A typical example of such processes is the sputtering process, whose surface evolution is described by the nonlinear stochastic

Kuramoto–Sivashinsky equation (KSE). In a simplified setting, the sputtering process includes two types of surface microprocesses, erosion and diffusion. The nonlinearity of the sputtering process originates from the dependence of the erosion rate on a nonlinear sputtering yield function [36]. In [101], feedback control of surface roughness in sputtering processes was designed based on a linearized stochastic KSE process model, which was identified using data from multiple kinetic Monte Carlo simulations of the same process. However, due to the fact that the inherent process nonlinearities are not explicitly considered in the linearized process model, it is expected that such a linear controller is only going to provide good closed-loop performance locally (i.e., for initial conditions close to the desired set point) for the nonlinear closed-loop system. To perform feedback control design for nonlinear stochastic processes, i.e., provide good performance for a wide range of process initial conditions and operating conditions, it is desirable that a nonlinear process model is directly used as the basis for controller synthesis. In this section, we use the stochastic KSE as an example to present a method for nonlinear control of nonlinear stochastic partial differential equations.

The stochastic KSE is a fourth-order nonlinear stochastic partial differential equation that describes the evolution of the height fluctuation for surfaces in a variety of material preparation processes including surface erosion by ion sputtering [36, 91], surface smoothing by energetic clusters [75] and ZrO_2 thin-film growth by reactive ion beam sputtering [124]. We consider the stochastic Kuramoto–Sivashinsky equation in a 1-dimensional domain [91] with distributed control in the spatial domain $[-\pi, \pi]$ (see also [8, 7, 30, 99] for a distributed control problem formulation for the deterministic KSE):

$$\frac{\partial h}{\partial t} = -\nu \frac{\partial^2 h}{\partial x^2} - \kappa \frac{\partial^4 h}{\partial x^4} + \frac{\lambda}{2} \left(\frac{\partial h}{\partial x} \right)^2 + \sum_{i=1}^{p} b_i(x) u_i(t) + \xi(x, t) \qquad (5.64)$$

subject to periodic boundary conditions

$$\frac{\partial^j h}{\partial x^j}(-\pi, t) = \frac{\partial^j h}{\partial x^j}(\pi, t), \qquad j = 0, \dots, 3, \qquad (5.65)$$

and with the initial condition

$$h(x, 0) = h_0(x), \qquad (5.66)$$

where ν, κ, and λ are parameters related to surface mechanisms [124], $x \in [-\pi, \pi]$ is the spatial coordinate, t is the time, $h(x, t)$ is the height of the surface at position x and time t, u_i is the ith manipulated input, p is the number of manipulated inputs, and b_i is the ith actuator distribution function [i.e., b_i determines how the control action computed by the ith control actuator, u_i, is distributed (e.g., point or distributed actuation) in the spatial interval $[-\pi, \pi]$]. $\xi(x, t)$ is a Gaussian noise with the following expressions for its mean and covariance:

$$\begin{aligned} \langle \xi(x, t) \rangle &= 0, \\ \langle \xi(x, t) \xi(x', t') \rangle &= \delta(x - x') \delta(t - t'), \end{aligned} \qquad (5.67)$$

where $\delta(\cdot)$ is the Dirac function, and $\langle \cdot \rangle$ denotes the expected value. Note that the noise covariance depends on both space x and time t.

Our objective is to control the expected roughness of the surface described by the stochastic KSE. The surface roughness, r, is defined in Eq. (5.36).

To study the dynamics of Eq. (5.64), we initially consider the eigenvalue problem of the linear operator in Eq. (5.64), which takes the form

$$A\bar{\phi}_n(x) = -\nu \frac{d^2 \bar{\phi}_n(x)}{dx^2} - \kappa \frac{d^4 \bar{\phi}_n(x)}{dx^4} = \lambda_n \bar{\phi}_n(x),$$

$$\frac{d^j \bar{\phi}_n}{dx^j}(-\pi) = \frac{d^j \bar{\phi}_n}{dx^j}(+\pi), \quad j = 0, \ldots, 3, \quad n = 1, \ldots, \infty, \qquad (5.68)$$

where λ_n denotes an eigenvalue and $\bar{\phi}_n$ denotes an eigenfunction. A direct computation of the solution of the above eigenvalue problem yields $\lambda_0 = 0$ with $\psi_0 = 1/\sqrt{2\pi}$, and $\lambda_n = \nu n^2 - \kappa n^4$ (λ_n is an eigenvalue of multiplicity two) with eigenfunctions $\phi_n = (1/\sqrt{\pi})\sin(nx)$ and $\psi_n = (1/\sqrt{\pi})\cos(nx)$ for $n = 1, \ldots, \infty$. Note that the $\bar{\phi}_n$ in Eq. (5.68) denotes either ϕ_n or ψ_n. From the expression of the eigenvalues, it follows that for fixed values of $\nu > 0$ and $\kappa > 0$, the number of unstable eigenvalues of the operator A in Eq. (5.68) is finite and the distance between two consecutive eigenvalues (i.e., λ_n and λ_{n+1}) increases as n increases.

To present the method that we use to control Eq. (5.64), we first derive nonlinear stochastic ODE approximations of Eq. (5.64) using Galerkin's method. To this end, we first expand the solution to Eq. (5.64) in an infinite series in terms of the eigenfunctions of the operator in Eq. (5.68) as follows:

$$h(x,t) = \sum_{n=1}^{\infty} \alpha_n(t)\phi_n(x) + \sum_{n=0}^{\infty} \beta_n(t)\psi_n(x), \qquad (5.69)$$

where $\alpha_n(t)$ and $\beta_n(t)$ are time-varying coefficients. Substituting the above expansion for the solution, $h(x,t)$, into Eq. (5.64) and taking the inner product with the adjoint eigenfunctions, $\phi_n^*(z) = (1/\sqrt{\pi})\sin(nz)$ and $\psi_n^*(z) = (1/\sqrt{\pi})\cos(nz)$, the following system of infinite nonlinear stochastic ODEs is obtained:

$$\frac{d\alpha_n}{dt} = (\nu n^2 - \kappa n^4)\alpha_n + f_{n\alpha} + \sum_{i=1}^{p} b_{i\alpha_n} u_i(t) + \xi_\alpha^n(t), \quad n = 1, \ldots, \infty,$$

$$\frac{d\beta_n}{dt} = (\nu n^2 - \kappa n^4)\beta_n + f_{n\beta} + \sum_{i=1}^{p} b_{i\beta_n} u_i(t) + \xi_\beta^n(t), \qquad (5.70)$$

where

$$f_{n\alpha} = \int_{-\pi}^{\pi} \phi_n^*(x) \left(\sum_{j=1}^{\infty} \alpha_j(t) \frac{d\phi_j}{dx}(x) + \sum_{j=0}^{\infty} \beta_j(t) \frac{d\psi_j}{dx}(x) \right)^2 dx,$$

$$\qquad (5.71)$$

$$f_{n\beta} = \int_{-\pi}^{\pi} \psi_n^*(x) \left(\sum_{j=1}^{\infty} \alpha_j(t) \frac{d\phi_j}{dx}(x) + \sum_{j=0}^{\infty} \beta_j(t) \frac{d\psi_j}{dx}(x) \right)^2 dx,$$

and

$$b_{i\alpha_n} = \int_{-\pi}^{\pi} \phi_n^*(x)b_i(x)dx,$$

$$b_{i\beta_n} = \int_{-\pi}^{\pi} \psi_n^*(x)b_i(x)dx,$$

$$\xi_\alpha^n(t) = \int_{-\pi}^{\pi} \xi(x,t)\phi_n^*(x)dx,$$

(5.72)

$$\xi_\beta^n(t) = \int_{-\pi}^{\pi} \xi(x,t)\psi_n^*(x)dx.$$

Using Result 4.1, we obtain $\langle \xi_\alpha^n(t)\xi_\alpha^n(t') \rangle = \delta(t - t')$ and $\langle \xi_\beta^n(t)\xi_\beta^n(t') \rangle = \delta(t - t')$. The controlled variable is the expected value of the square of the surface roughness defined in Eq. (5.36), $\langle r^2(t) \rangle$. According to Eq. (5.69), we have $\bar{h}(t) = \beta_0(t)\psi_0$. Therefore, $\langle r^2(t) \rangle$ can be rewritten in terms of $\alpha_n(t)$ and $\beta_n(t)$ as follows:

$$\langle r^2(t) \rangle = \frac{1}{2\pi} \left\langle \int_{-\pi}^{\pi} (h(x,t) - \bar{h}(t))^2 dx \right\rangle$$

$$= \frac{1}{2\pi} \left\langle \int_{-\pi}^{\pi} \left[\sum_{i=1}^{\infty} \alpha_i(t)\phi_i(x) + \sum_{i=0}^{\infty} \beta_i(t)\psi_i(x) - \beta_0(t)\psi_0 \right]^2 dx \right\rangle$$

(5.73)

$$= \frac{1}{2\pi} \left\langle \int_{-\pi}^{\pi} \sum_{i=1}^{\infty} \left[\alpha_i^2(t)\phi_i^2(x) + \beta_i^2(t)\psi_i^2(x) \right] dx \right\rangle$$

$$= \frac{1}{2\pi} \left\langle \sum_{i=1}^{\infty} (\alpha_i^2(t) + \beta_i^2(t)) \right\rangle = \frac{1}{2\pi} \sum_{i=1}^{\infty} \left[\langle \alpha_i^2(t) \rangle + \langle \beta_i^2(t) \rangle \right].$$

The surface roughness control problem for the stochastic KSE in Eq. (5.64) is formulated as that of controlling the covariance of the states α_n and β_n in the nonlinear stochastic ODE system in Eq. (5.70).

Remark 5.5. Note that in the stochastic KSE in Eq. (5.64), the covariance of $\xi(x,t)$ is normalized to be $\delta(x - x')\delta(t - t')$. In general, the covariance of the noise term of the stochastic KSE is $\varsigma^2\delta(x - x')\delta(t - t')$, where ς^2 is a process parameter derived based on the rates of surface microscopic processes [36]. The normalization procedure is detailed in the simulation section of this work. This convenience is adopted to simplify our presentation.

5.4.1 Model Reduction

Due to its infinite-dimensional nature, the system in Eq. (5.70) cannot be directly used for the design of controllers that can be implemented in practice (i.e., the practical implementation of controllers that are designed on the basis of this system will require the computation of infinite sums, which cannot be done by a computer). Instead, we will base the controller design on a finite-dimensional approximation of this system. Subsequently, we will show that the resulting controller will enforce the desired control objective in the closed-loop infinite-dimensional system. Specifically, we rewrite the system in Eq. (5.70) as follows:

$$\frac{dx_s}{dt} = \Lambda_s x_s + f_s(x_s, x_f) + B_s u + \xi_s,$$

$$\frac{dx_f}{dt} = \Lambda_f x_f + f_f(x_s, x_f) + B_f u + \xi_f,$$

(5.74)

where $x_s = [\alpha_1 \cdots \alpha_m \, \beta_1 \cdots \beta_m]^T$, $x_f = [\alpha_{m+1} \, \beta_{m+1} \cdots \,]^T$, $\Lambda_s = \text{diag}[\lambda_1 \cdots \lambda_m \, \lambda_1 \cdots \lambda_m]$, $\Lambda_f = \text{diag}[\lambda_{m+1} \, \lambda_{m+1} \, \lambda_{m+2} \, \lambda_{m+2} \cdots]$, $f_s(x_s, x_f) = [f_{1\alpha}(x_s, x_f) \cdots f_{m\alpha}(x_s, x_f) \, f_{1\beta}(x_s, x_f) \cdots f_{m\beta}(x_s, x_f)]^T$, $f_f(x_s, x_f) = [f_{(m+1)\alpha}(x_s, x_f) \, f_{(m+1)\beta}(x_s, x_f) \cdots \,]^T$, $u = [u_1 \cdots u_p]^T$, $\xi_s = [\xi_\alpha^1 \cdots \xi_\alpha^m \, \xi_\beta^1 \cdots \xi_\beta^m]^T$, and $\xi_f = [\xi_\alpha^{m+1} \, \xi_\beta^{m+1} \cdots]^T$.

$$
B_s = \begin{bmatrix} b_{1_{\alpha_1}} & \cdots & b_{p_{\alpha_1}} \\ \vdots & \ddots & \vdots \\ b_{1_{\alpha_m}} & \cdots & b_{p_{\alpha_m}} \\ b_{1_{\beta_1}} & \cdots & b_{p_{\beta_1}} \\ \vdots & \ddots & \vdots \\ b_{1_{\beta_m}} & \cdots & b_{p_{\beta_m}} \end{bmatrix}, \qquad
B_f = \begin{bmatrix} b_{1_{\alpha_{m+1}}} & \cdots & b_{p_{\alpha_{m+1}}} \\ b_{1_{\beta_{m+1}}} & \cdots & b_{p_{\beta_{m+1}}} \\ b_{1_{\alpha_{m+2}}} & \cdots & b_{p_{\alpha_{m+2}}} \\ b_{1_{\beta_{m+2}}} & \cdots & b_{p_{\beta_{m+2}}} \\ \vdots & \vdots & \vdots \end{bmatrix}.
$$

(5.75)

In our development, we will need the following notations. The 2-norms for the vectors x_s, x_f, and $f_s(x_s, x_f)$ are defined as follows:

$$\|x_s\|_2 = \sqrt{\sum_{j=1}^{m} \alpha_j^2 + \sum_{j=1}^{m} \beta_j^2},$$

$$\|x_f\|_2 = \sqrt{\sum_{j=m+1}^{\infty} \alpha_j^2 + \sum_{j=m+1}^{\infty} \beta_j^2},$$

(5.76)

$$\|f_s(x_s, x_f)\|_2 = \sqrt{\sum_{j=1}^{m} f_{j\alpha}^2(x_s, x_f) + \sum_{j=1}^{m} f_{j\beta}^2(x_s, x_f)}.$$

The covariance matrices for $x_s(t)$, $x_f(t)$, $P_s(t)$ and $P_f(t)$, are defined as follows:

$$P_s(t) = \langle x_s(t) x_s^T(t) \rangle, \qquad P_f(t) = \langle x_f(t) x_f^T(t), \rangle \qquad (5.77)$$

where $\langle \cdot \rangle$ denotes the expected value and $x_s^T(t)$ and $x_f^T(t)$ are transposes of the vectors $x_s(t)$ and $x_f(t)$, respectively.

We note that the subsystem x_f in Eq. (5.74) is infinite-dimensional. Neglecting the x_f subsystem, the following $2m$-dimensional system is obtained:

$$\frac{d\tilde{x}_s}{dt} = \Lambda_s \tilde{x}_s + f_s(\tilde{x}_s, 0) + B_s u + \xi_s, \qquad (5.78)$$

where the tilde symbol in $\tilde{x}_s$ denotes that this state variable is associated with a finite-dimensional system.

5.4.2 Feedback Control Design

We design the nonlinear state feedback controller on the basis of Eq. (5.78). To simplify our development, we assume that $p = 2m$ (i.e., the number of control actuators is equal to the dimension of the finite-dimensional system) and pick the actuator distribution functions such that B_s^{-1} exists. The state feedback control law then takes the form

$$u = B_s^{-1} \{ (\Lambda_{cs} - \Lambda_s) \tilde{x}_s - f_s(\tilde{x}_s, 0) \}, \qquad (5.79)$$

where the matrix Λ_{cs} contains the desired poles of the closed-loop system; $\Lambda_{cs} = \mathrm{diag}[\lambda_{c\alpha 1} \cdots \lambda_{c\alpha m} \, \lambda_{c\beta 1} \cdots \lambda_{c\beta m}]$, and $\lambda_{c\alpha i}$ and $\lambda_{c\beta i}$ $(1 \leq i \leq m)$ are desired poles of the closed-loop finite-dimensional system, which satisfy $\mathrm{Re}\{\lambda_{c\alpha i}\} < 0$ and $\mathrm{Re}\{\lambda_{c\beta i}\} < 0$ for $(1 \leq i \leq m)$ and can be determined from the desired closed-loop surface roughness level. The method to determine the eigenvalues of Λ_{cs} will be discussed in Section 5.4.3.

The control action is computed using the formula in Eq. (5.79), and the computation as cost is proportional to the number of actuators, p. Since B_s^{-1} depends only on the configuration of the control actuators, it can be computed offline. The major computational requirement involved in Eq. (5.79) is the evaluation of the nonlinear term $f_s(\tilde{x}_s, 0)$, whose specific form is given in Eq. (5.80):

$$f_s(\tilde{x}_s, 0) = [f_{1\alpha}(\tilde{x}_s, 0) \cdots f_{m\alpha}(\tilde{x}_s, 0) f_{1\beta}(\tilde{x}_s, 0) \cdots f_{m\beta}(\tilde{x}_s, 0)]^T,$$

$$f_{n\alpha}(\tilde{x}_s, 0) = \int_{-\pi}^{\pi} \phi_n^*(x) \left(\sum_{j=1}^{m} \alpha_j(t) \frac{d\phi_j}{dx}(x) + \sum_{j=1}^{m} \beta_j(t) \frac{d\psi_j}{dx}(x) \right)^2 dx, \qquad (5.80)$$

$$f_{n\beta}(\tilde{x}_s, 0) = \int_{-\pi}^{\pi} \psi_n^*(x) \left(\sum_{j=1}^{m} \alpha_j(t) \frac{d\phi_j}{dx}(x) + \sum_{j=1}^{m} \beta_j(t) \frac{d\psi_j}{dx}(x) \right)^2 dx,$$

where $n = 1, \ldots, m$. Therefore, the computation of $f_s(\tilde{x}_s, 0)$ involves standard numerical operations and can be completed very fast relative to the time scale of process evolution using currently available computing power.

We will demonstrate in Section 5.4.3 that the expected surface roughness of the closed-loop infinite-dimensional system in Eq. (5.70) can be controlled to the desired level by using the state feedback controller in Eq. (5.79), which uses only a finite number of actuators.

5.4.3 Analysis of the Closed-Loop Infinite-Dimensional System

By applying the controller in Eq. (5.79) to the infinite-dimensional system in Eq. (5.74), and using that $\epsilon = |\lambda_1|/|\lambda_{m+1}|$, the closed-loop system takes the form

$$
\begin{aligned}
\frac{dx_s}{dt} &= \Lambda_{cs} x_s + (f_s(x_s, x_f) - f_s(x_s, 0)) + \xi_s, \\
\epsilon \frac{dx_f}{dt} &= \Lambda_{f\epsilon} x_f + \epsilon B_f B_s^{-1} (\Lambda_{cs} - \Lambda_s) \tilde{x}_s \\
&\quad + \epsilon f_f(x_s, x_f) - \epsilon B_f B_s^{-1} f_s(\tilde{x}_s, 0) + \epsilon \xi_f,
\end{aligned}
\tag{5.81}
$$

where λ_1 and λ_{m+1} are the first and the $(m+1)$th eigenvalues of the linear operator in Eq. (5.68), and $\Lambda_{f\epsilon} = \text{diag}[\lambda_{\epsilon 1}\ \lambda_{\epsilon 1}\ \lambda_{\epsilon 2}\ \lambda_{\epsilon 2} \cdots]$ is an infinite-dimensional matrix defined as $\Lambda_{f\epsilon} = \epsilon \cdot \Lambda_f$.

Computing the linearization of the nonlinear system in Eq. (5.81) around $(x_s, x_f) = (0,0)$ and using the fact that the terms $\{f_s(x_s, x_f) - f_s(x_s, 0)\}$, $f_f(x_s, x_f)$, and $f_s(x_s, 0)$ include second-order terms and do not include linear terms [this follows from the quadratic structure of the nonlinear term of the stochastic KSE and from Eq. (5.71)], we obtain the following linear system:

$$
\begin{aligned}
\frac{dx_s}{dt} &= \Lambda_{cs} x_s + \xi_s, \\
\epsilon \frac{dx_f}{dt} &= \Lambda_{f\epsilon} x_f + \epsilon B_f B_s^{-1} (\Lambda_{cs} - \Lambda_s) \tilde{x}_s + \epsilon \xi_f.
\end{aligned}
\tag{5.82}
$$

Due to the stability properties of Λ_{cs} and $\Lambda_{f\epsilon}$ and the decoupled nature of the system in Eq. (5.82), this system is asymptotically stable. Thus, the nonlinear system in Eq. (5.81) is locally (i.e., for sufficiently small initial conditions) asymptotically stable. This implies that under the assumption that the initial condition is sufficiely small, as $t \to \infty$, the covariance matrices of x_s and x_f of the system in Eq. (5.81) converge to $P_s(\infty)$ and $P_f(\infty)$, respectively. $P_s(\infty)$ and $P_f(\infty)$ are defined as follows:

$$
P_s(\infty) = \lim_{t\to\infty} \langle x_s(t) x_s^T(t) \rangle, \qquad P_f(\infty) = \lim_{t\to\infty} \langle x_f(t) x_f^T(t) \rangle,
\tag{5.83}
$$

where $\langle \cdot \rangle$ denotes the expected value, and $x_s^T(t)$ and $x_f^T(t)$ are transposes of the vectors $x_s(t)$ and $x_f(t)$, respectively.

We now proceed to characterize the accuracy with which the closed-loop surface roughness is controlled. Theorem 5.1 provides estimates of the contribution of the expected surface roughness from the x_s and x_f subsystems of the closed-loop system in Eq. (5.81) and a characterization of the expected value of the surface roughness enforced by the controller in Eq. (5.79) in the closed-loop stochastic KSE. The proof of Theorem 5.1 is also given below.

Theorem 5.1. *Consider the closed-loop stochastic KSE in Eq. (5.81). Define the expected surface roughness and the contribution to the expected surface roughness of the closed-loop system from the x_f and x_s subsystems as $t \to \infty$ as follows:*

$$\langle r^2(\infty) \rangle = \frac{1}{2\pi} \sum_{i=1}^{\infty} \left[\langle \alpha_i^2(\infty) \rangle + \langle \beta_i^2(\infty) \rangle \right],$$

$$\langle r_f^2(\infty) \rangle = \frac{1}{2\pi} \sum_{i=m+1}^{\infty} \left[\langle \alpha_i^2(\infty) \rangle + \langle \beta_i^2(\infty) \rangle \right], \tag{5.84}$$

$$\langle r_s^2(\infty) \rangle = \frac{1}{2\pi} \sum_{i=1}^{m} \left[\langle \alpha_i^2(\infty) \rangle + \langle \beta_i^2(\infty) \rangle \right],$$

where $\langle \cdot \rangle$ denotes the expected value, $\langle r^2(\infty) \rangle$ is the expected surface roughness of the closed-loop system in Eq. (5.81), $\langle r_f^2(\infty) \rangle$ is the contribution to the expected surface roughness from the x_f subsystem in Eq. (5.81), $\langle r_s^2(\infty) \rangle$ is the contribution to the expected surface roughness from the x_s subsystem in Eq. (5.81), $x_f = [\alpha_{m+1} \ \beta_{m+1} \ \alpha_{m+2} \ \beta_{m+2} \ \cdots \ ]^T$, and $x_s = [\alpha_1 \ \cdots \ \alpha_m \ \beta_1 \ \cdots \ \beta_m]^T$.
Then, there exist $\mu^* > 0$ and $\epsilon^* > 0$ such that if $\|x_{f0}\|_2 + \|x_{s0}\|_2 \le \mu^*$ and $\epsilon \in (0, \epsilon^*]$, then $\langle r_f^2(\infty) \rangle$, $\langle r_s^2\infty) \rangle$, and $\langle r^2(\infty) \rangle$ satisfy

$$\langle r_f^2(\infty) \rangle = O(\epsilon), \tag{5.85}$$

$$\langle r_s^2(\infty) \rangle = \frac{1}{4\pi} \sum_{i=1}^{m} \left(\frac{1}{|\lambda_{c\alpha i}|} + \frac{1}{|\lambda_{c\beta i}|} \right) + O(\sqrt{\epsilon}), \tag{5.86}$$

$$\langle r^2(\infty) \rangle = \frac{1}{4\pi} \sum_{i=1}^{m} \left(\frac{1}{|\lambda_{c\alpha i}|} + \frac{1}{|\lambda_{c\beta i}|} \right) + O(\sqrt{\epsilon}), \tag{5.87}$$

where x_{f0} and x_{s0} are the initial conditions for x_f and x_s in Eq. (5.81), respectively, and $\lambda_{c\alpha i}$, $\lambda_{c\beta i}$ ($i = 1, 2, \ldots, m$) are the eigenvalues of Λ_{cs} in the system in Eq. (5.81).

Proof of Theorem 5.1. The proof of Theorem 5.1 includes three parts. First, we compute the contribution to the expected surface roughness from the x_f subsystem in Eq. (5.81) and prove Eq. (5.85) in Theorem 5.1. Then, we compute the contribution to the expected surface roughness from the x_s subsystem in Eq. (5.81) and prove Eq. (5.86) in Theorem 5.1. Finally, the proof of Theorem 5.1 is completed by proving Eq. (5.87) based on the results in Eqs. (5.85) and (5.86). Since we work with sufficiently small initial conditions, local stability of the closed-loop nonlinear infinite-dimensional system can be proved by using the linearization argument of Section 5.4.3 and is used without further proof in the remainder.

Proof of Eq. (5.85) in Theorem 5.1. Consider the closed-loop system in Eq. (5.81) and note that the terms on the right-hand side of the x_f subsystem constitute an $O(\epsilon)$ approximation to the term $\Lambda_{f\epsilon}x_f$. Consider also the following linear system:

$$\epsilon\frac{\partial \bar{x}_f}{\partial t} = \Lambda_{f\epsilon}\bar{x}_f + \epsilon\xi_f. \tag{5.88}$$

Following a similar approach to the one employed in the proof of Theorem A.1 in [87, p. 361], we have that there exists an $\hat{\epsilon}^* > 0$ such that for all $\epsilon \in (0, \hat{\epsilon}^*]$, we have that

$$x_f(t) = \bar{x}_f(t) + O(\sqrt{\epsilon}). \tag{5.89}$$

Therefore, we have the following estimate for $\langle \|x_f(t)\|_2^2 \rangle$:

$$\langle \|x_f(t)\|_2^2 \rangle = \langle \|\bar{x}_f(t) + O(\sqrt{\epsilon})\|_2^2 \rangle \le 2\langle \|\bar{x}_f(t)\|_2^2 \rangle + O(\epsilon). \tag{5.90}$$

Furthermore, $\langle \|x_f(t)\|_2^2 \rangle$ and $\langle \|\bar{x}_f(t)\|_2^2 \rangle$ are equal to the traces of the covariance matrices of $x_f(t)$ and $\bar{x}_f(t)$, $P_f(t) = \langle x_f(t)x_f(t)^T \rangle$ and $\bar{P}_f(t) = \langle \bar{x}_f(t)\bar{x}_f(t)^T \rangle$, respectively. Finally, as $t \to \infty$, $P_f(t)$ and $\bar{P}_f(t)$ converge to $P_f(\infty)$ and $\bar{P}_f(\infty)$, respectively [both $P_f(\infty)$ and $\bar{P}_f(\infty)$ are bounded quantities, which follows from closed-loop stability]. We note that the 2-norm of x_f, $\|x_f\|_2$, is defined in Eq. (5.76) and $P_f(\infty)$ is defined in Eq. (5.83). Because $\Lambda_{f\epsilon}$ is a diagonal matrix, the trace of matrix $\bar{P}_f$ can be computed as follows [101]:

$$\text{Tr}\{\bar{P}_f\} = \frac{\epsilon}{2} \cdot \sum_{i=1}^{\infty} \left|\frac{1}{\lambda_{\epsilon i}}\right|, \tag{5.91}$$

where $\lambda_{\epsilon i}$ ($i = 1, 2, \ldots, \infty$) are the eigenvalues of the matrix $\Lambda_{f\epsilon}$ in Eq. (5.88). Specifically, $|\lambda_{\epsilon 1}| = |\nu - \kappa|$ and

$$|\lambda_{\epsilon i}| = \left|\frac{(\nu - \kappa)\left[\nu(2m+i)^2 - \kappa(2m+i)^4\right]}{\nu(2m+1)^2 - \kappa(2m+1)^4}\right|, \qquad i = 1, 2, \ldots.$$

It is clear that $|\lambda_{\epsilon i}|$ increases with respect to i in the order of $(2m+i)^4$, where $2m$ is the size of the x_s subsystem in Eq. (5.81). Therefore, $\sum_{i=1}^{\infty}\left|\frac{1}{\lambda_{\epsilon i}}\right|$ converges to a finite positive number, and, thus, there exists a positive real number $k_{f\epsilon}$ such that

$$\text{Tr}\{\bar{P}_f\} < \frac{\epsilon}{2} \cdot k_{f\epsilon}. \tag{5.92}$$

Therefore, it follows that

$$\text{Tr}\{\bar{P}_f\} = \langle \|\bar{x}_f(\infty)\|_2^2 \rangle = O(\epsilon). \tag{5.93}$$

According to Eq. (5.90), it follows that the contribution to the expected surface roughness from x_f is $O(\epsilon)$, i.e.

$$\langle r_f^2(\infty)\rangle = \frac{1}{2\pi}\langle\|x_f(\infty)\|_2^2\rangle = \frac{1}{2\pi}\sum_{i=m+1}^{\infty}\left[\langle\alpha_i^2(\infty)\rangle + \langle\beta_i^2(\infty)\rangle\right] = O(\epsilon). \quad (5.94)$$

This completes the proof of Eq. (5.85) in Theorem 5.1. □

Proof of Eq. (5.86) in Theorem 5.1. Consider the x_s subsystem of the closed-loop system in Eq. (5.81). First, we note that there exists a positive real number k_{1s} such that [32, 28]

$$\|f_s(x_s, x_f) - f_s(x_s, 0)\|_2 < k_{1s}\|x_f\|_2, \quad (5.95)$$

where the definitions of $\|x_f\|_2$ and $\|f_s(\cdot)\|_2$ can be found in Eq. (5.76). From Eq. (5.89), we have the following estimate for $\|x_f\|_2$ for $t \geq t_b$ (where t_b is the time needed for $\|\bar{x}_f(t)\|$ to approach zero and $t_b \to 0$ as $\epsilon \to 0$):

$$\|x_f(t)\|_2 = \sqrt{\sum_{i=m+1}^{\infty}\left[\langle\alpha_i^2(\infty)\rangle + \langle\beta_i^2(\infty)\rangle\right]} = O(\sqrt{\epsilon}). \quad (5.96)$$

This implies that we have the following estimate for $\|f_s(x_s, x_f) - f_s(x_s, 0)\|_2$ for $t \geq t_b$:

$$\|f_s(x_s, x_f) - f_s(x_s, 0)\|_2 = O(\sqrt{\epsilon}). \quad (5.97)$$

Therefore, the solution of the following system consists of an $O(\sqrt{\epsilon})$ approximation of the x_s of Eq. (5.81) [87, Theorem A.1, p. 361]:

$$\frac{d\bar{x}_s}{dt} = \Lambda_{cs}\bar{x}_s + \xi_s. \quad (5.98)$$

In particular, there exists an $\hat{\epsilon}^{**} > 0$ such that for all $\epsilon \in (0, \hat{\epsilon}^{**}]$, it holds that

$$x_s(t) - \bar{x}_s(t) = O(\sqrt{\epsilon}) \quad (5.99)$$

and

$$\|x_s(t)\|_2^2 - \|\bar{x}_s(t)\|_2^2 = (\|x_s(t)\|_2 - \|\bar{x}_s(t)\|_2)\cdot(\|x_s(t)\|_2 + \|\bar{x}_s(t)\|_2) = O(\sqrt{\epsilon}). \quad (5.100)$$

Because $\|x_s(t)\|_2$ and $\|\bar{x}_s(t)\|_2$ are bounded for all $t > 0$, $\langle\|x_s(t)\|_2^2\rangle$ and $\langle\|\bar{x}_s(t)\|_2^2\rangle$ are equal to the traces of the covariance matrices of $x_s(t)$ and $\bar{x}_s(t)$, $P_s(t) = \langle x_s(t)x_s(t)^T\rangle$ and $\bar{P}_s(t) = \langle\bar{x}_s(t)\bar{x}_s(t)^T\rangle$, respectively. As $t \to \infty$, $P_s(t)$ and $\bar{P}_s(t)$ converge to $P_s(\infty)$ and $\bar{P}_s(\infty)$, respectively. The definition of $P_s(\infty)$ can be found in Eq. (5.83).

Because Λ_{cs} is a diagonal matrix, the trace of matrix $\bar{P}_s(\infty)$ can be computed as follows [101]:

$$\text{Tr}\{\bar{P}_s(\infty)\} = \frac{1}{2}\cdot\sum_{i=1}^{m}\left(\left|\frac{1}{\lambda_{c\alpha i}}\right| + \left|\frac{1}{\lambda_{c\beta i}}\right|\right), \quad (5.101)$$

where $\lambda_{c\alpha i}$ and $\lambda_{c\beta i}$ $(i = 1, 2, \ldots, m)$ are the eigenvalues of the matrix Λ_{cs} in Eq. (5.81).

Therefore, it holds that

$$\langle\|\bar{x}_s(\infty)\|_2^2\rangle = Tr\{\bar{P}_s(\infty)\} = \frac{1}{2}\cdot\sum_{i=1}^{m}\left(\left|\frac{1}{\lambda_{c\alpha i}}\right| + \left|\frac{1}{\lambda_{c\beta i}}\right|\right). \tag{5.102}$$

According to Eq. (5.100), it holds that the contribution to the expected surface roughness from x_s is as follows:

$$\langle r_s^2(\infty)\rangle = \frac{1}{2\pi}\langle\|x_s(\infty)\|_2^2\rangle = \frac{1}{2\pi}\sum_{i=1}^{m}\left[\langle\alpha_i^2(\infty)\rangle + \langle\beta_i^2(\infty)\rangle\right]$$

$$= \frac{1}{4\pi}\sum_{i=1}^{m}\left(\left|\frac{1}{\lambda_{c\alpha i}}\right| + \left|\frac{1}{\lambda_{c\beta i}}\right|\right) + O(\sqrt{\epsilon}). \tag{5.103}$$

This completes the proof of Eq. (5.86) in Theorem 5.1.   □

Proof of Eq. (5.87) in Theorem 5.1. The expected surface roughness from the closed-loop system $\langle r^2(\infty)\rangle$ includes contributions from both the x_s subsystem and the x_f subsystem in Eq. (5.81). Therefore, we have the following equation for $\langle r^2(\infty)\rangle$:

$$\langle r^2(\infty)\rangle = \langle r_s^2(\infty)\rangle + \langle r_f^2(\infty)\rangle. \tag{5.104}$$

Using Eqs. (5.85) and (5.86), we immediately have

$$\langle r^2(\infty)\rangle = \frac{1}{4\pi}\sum_{i=1}^{m}\left(\frac{1}{|\lambda_{c\alpha i}|} + \frac{1}{|\lambda_{c\beta i}|}\right) + O(\sqrt{\epsilon}) + O(\epsilon), \tag{5.105}$$

since as $\epsilon\to 0$, it holds that

$$\frac{O(\epsilon)}{O(\sqrt{\epsilon})}\to 0. \tag{5.106}$$

The $O(\epsilon)$ term in Eq. (5.105) is negligible and there exists an $\epsilon^* = \min(\hat{\epsilon}^*, \hat{\epsilon}^{**})$ such that if $\epsilon\in(0, \epsilon^*]$, then

$$\langle r^2(\infty)\rangle = \frac{1}{4\pi}\sum_{i=1}^{m}\left(\frac{1}{|\lambda_{c\alpha i}|} + \frac{1}{|\lambda_{c\beta i}|}\right) + O(\sqrt{\epsilon}), \tag{5.107}$$

where $\lambda_{c\alpha i}, \lambda_{c\beta i}$ $(i = 1, 2, \ldots, m)$ are eigenvalues of Λ_{cs} in the system in Eq. (5.81).
     This completes the proof of Theorem 5.1.   ■

Remark 5.6. Note that to control the expected value of the surface roughness to $\langle r^2(\infty)\rangle$, we need to design a controller to assign the eigenvalues of the matrix Λ_{cs} in the system in Eq. (5.81) to appropriate values. The controller that assigns the eigenvalues of the matrix Λ_{cs} in the system in Eq. (5.81) to satisfy Eq. (5.87) is not unique. Consequently, for a fixed number of actuators, p, the controller that can drive the closed-loop surface roughness to a desired level is not unique either. Furthermore, we note that the proposed nonlinear feedback controller in Eq. (5.79) is a multivariable controller (i.e., the numbers of the manipulated inputs adjusted by this controller is equal to p). Therefore, the number of independent output variables that this controller is capable of simultaneously regulating is equal to p. If control of surface configuration variables other than the surface roughness is of interest (for example, surface coverage, island size, etc.), then these variables should be mathematically expressed as controlled outputs of the stochastic PDE and the nonlinear feedback controller should be designed to regulate these new outputs to the desired set-point values in a similar way to the one that is followed to achieve this task for the expected value of the surface roughness.

Remark 5.7. In case where the desired value of the steady-state surface roughness of the closed-loop system is $\langle r_d^2 \rangle$, the controller should be designed such that

$$\langle r_d^2 \rangle = \frac{1}{4\pi} \sum_{i=1}^{m} \left(\frac{1}{|\lambda_{c\alpha i}|} + \frac{1}{|\lambda_{c\beta i}|} \right).$$

Under this controller, the expected surface roughness of the infinite-dimensional system is shown in Eq. (5.87), which is an $O(\sqrt{\epsilon})$ approximation of $\langle r_d^2 \rangle$, which means that there exists a positive real number k_r such that $|\langle r^2(\infty) \rangle - \langle r_d^2 \rangle| < k_r \cdot \sqrt{\epsilon}$. Under the assumption that the number of control actuators is equal to the dimension of the x_s subsystem, the value of ϵ is dependent on the number of actuators used by the controller. Therefore, the larger the number of control actuators used, the smaller the ϵ. Consequently, the closed-loop surface roughness is closer to the desired surface roughness as the number of control actuators used to control the process increases. If the allowable error between the closed-loop surface roughness and the desired surface roughness is prespecified as $e_r = |\langle r_d^2 \rangle - \langle r^2(\infty) \rangle|$, then the number of control actuators should be chosen such that $k_r \cdot \sqrt{\epsilon} < e_r$ to achieve the desired closed-loop performance. However, it is not straightforward to solve for k_r analytically. Therefore, the number of control actuators can be determined using a two-step procedure. First, an estimate of the number of actuators, p_1, is made and an ϵ_1 is computed. Closed-loop simulations can be performed to evaluate the error between the expected closed-loop surface roughness and the desired value, $\bar{e}_{r1}$, when p_1 actuators are used. If $\bar{e}_{r1} > e_r$, then the number of actuators should be increased to p_2 (and the value of ϵ is reduced from ϵ_1 to ϵ_2) such that $\bar{e}_{r1} \cdot \sqrt{\epsilon_2} > e_r \cdot \sqrt{\epsilon_1}$ in order to achieve the desired closed-loop performance.

Remark 5.8. We note that a full-scale model of a sputtering process would consist of a two-dimensional lattice representation of the surface. Although we developed the method for nonlinear feedback control design based on a one-dimensional lattice representation of the surface, it is possible to extend the proposed method to control the surface roughness of material preparation processes taking place in two-dimensional domains. In a two-dimensional in space process, the feedback control design and the analysis of the closed-loop system will be based on a two-dimensional extension of the model in Eq. (5.74). Moreover, Eq. (5.74) will be obtained by solving the eigenvalue/eigenfunction problem of the operator A in the two-dimensional spatial domain subject to the appropriate boundary conditions; this can be achieved in a similar way to that followed for the one-dimensional spatial domain (see Section 4.3.2 for results on the solution of the eigenvalue/eigenfunction problem for a two-dimensional spatial domain). Once the modal representation in Eq. (5.74) corresponding to the two-dimensional PDE is obtained, the method for control design and closed-loop analysis presented above can be applied to control the surface roughness.

5.4.4 Application to the Stochastic Kuramoto-Sivashinsky Equation

In this subsection, we consider the following stochastic KSE with spatially distributed control:

$$\frac{\partial h}{\partial t} = -\nu \frac{\partial^2 h}{\partial x^2} - \kappa \frac{\partial^4 h}{\partial x^4} + \frac{\lambda}{2} \left(\frac{\partial h}{\partial x} \right)^2 + \sum_{i=1}^{p} \hat{b}_i(x) u_i(t) + \xi'(x, t), \qquad (5.108)$$

where u_i is the ith manipulated input, p is the number of manipulated inputs, $\hat{b}_i$ is the ith actuator distribution function [i.e., $\hat{b}_i$ determines how the control action computed by the ith control actuator, u_i, is distributed (e.g., point or distributed actuation) in the spatial interval $[-\pi, \pi]$], $\nu = 1.975 \times 10^{-4}$, $\kappa = 1.58 \times 10^{-4}$, $\lambda = 1.975 \times 10^{-4}$, $x \in [-\pi, \pi]$ is the spatial coordinate, t is the time, $h(x, t)$ is the height of the surface at position x and time t, and $\xi'(x, t)$ is a Gaussian noise with zero mean and covariance:

$$\langle \xi'(x, t) \xi'(x', t') \rangle = \varsigma^2 \delta(x - x') \delta(t - t'), \qquad (5.109)$$

where $\varsigma = 3.52 \times 10^{-5}$. Note that the difference between Eq. (5.64) and Eq. (5.108) is that in Eq. (5.108), the covariance of the Gaussian noise, $\xi'(x, t)$, is $\varsigma^2 \delta(x - x') \delta(t - t')$, while in Eq. (5.64), the covariance of the Gaussian noise, $\xi(x, t)$, is $\delta(x - x') \delta(t - t')$.

We normalize the covariance of the Gaussian noise in the system in Eq. (5.108) to be $\delta(x - x') \delta(t - t')$ by introducing a new variable for the height of the surface, $h'(x, t) = h(x, t)/\varsigma$, and new actuator distribution functions, $b_i(x) = \hat{b}_i(x)/\varsigma$, for $i = 1, \ldots, p$. Equation (5.108), therefore, can be rewritten as follows:

$$\frac{\partial h'}{\partial t} = -\nu \frac{\partial^2 h'}{\partial x^2} - \kappa \frac{\partial^4 h'}{\partial x^4} + \frac{\lambda}{2} \left(\frac{\partial h'}{\partial x} \right)^2 + \sum_{i=1}^{p} b_i(x) u_i(t) + \xi(x, t), \qquad (5.110)$$

where $\xi(x, t) = \xi'(x, t)/\varsigma$ and $\langle \xi(x, t) \xi(x', t') \rangle = \delta(x - x') \delta(t - t')$. Therefore, the system in Eq. (5.110) is consistent with the system in Eq. (5.64) which is the basis for feedback control design and closed-loop analysis. We also use the system in Eq. (5.110) for all simulations in this work.

A 200th-order stochastic ordinary differential equation approximation of Eq. (5.110) obtained via Galerkin's method is used to simulate the process (the use of higher-order approximations led to identical numerical results, thereby implying that the following simulation runs are independent of the discretization). The δ function involved in the covariances of ξ_α^n and ξ_β^n is approximated by $1/\Delta t$, where Δt is the integration time step.

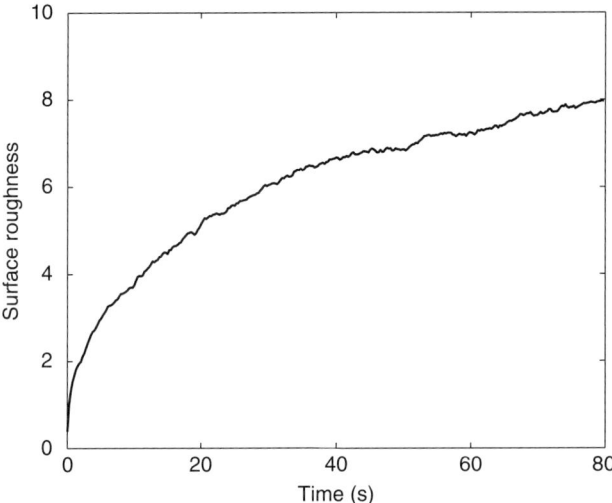

Fig. 5.12. The open-loop profile of the expected surface roughness resulting from the computation of the average of 100 independent simulation runs of the stochastic KSE in Eq. (5.110).

Open-Loop Dynamics of the Stochastic KSE

In the first simulation, we compute the expected value of the open-loop surface roughness profile from the solution of the stochastic KSE in Eq. (5.110) by setting $u_i(t) = 0$ for $i = 1, \ldots, p$. For $\nu = 1.975 \times 10^{-4}$ and $\kappa = 1.58 \times 10^{-4}$, the stochastic KSE possesses one positive eigenvalue. Therefore, the zero solution of the open-loop system is unstable. Surface roughness profiles obtained from 100 independent simulation runs using the same parameters are averaged and the resulting expected surface roughness profile is shown in Fig. 5.12. The value of the open-loop surface roughness increases due to the open-loop instability of the zero solution but remains bounded due to the bounding nature of the nonlinear terms in the stochastic KSE.

Closed-Loop Simulation of the Stochastic KSE Under Nonlinear Control

In the closed-loop simulation under nonlinear control, we design a nonlinear state feedback controller based on a 20th-order stochastic ODE approximation constructed by using the first 20 eigenmodes of the system in Eq. (5.70) and apply this controller to a 200th-order approximation of the nonlinear stochastic KSE. Twenty control actuators are used to control the system. The ith actuator distribution function is taken to be

$$
b_i(z) = \begin{cases} \dfrac{1}{\sqrt{\pi}} \sin(iz), & i = 1, \ldots, 10, \\[2ex] \dfrac{1}{\sqrt{\pi}} \cos[(i - 10)z], & i = 11, \ldots, 20. \end{cases} \tag{5.111}
$$

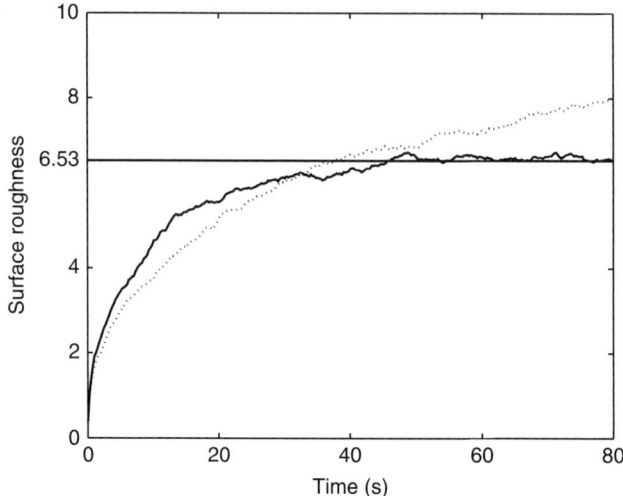

Fig. 5.13. The closed-loop profile of the expected value of the surface roughness (solid line) vs. the open-loop profile of the expected value of the surface roughness (dotted line) when the controller is designed based on the first 20 modes.

Under this control problem formulation, $m = 10$ and the value of $\epsilon = |\lambda_1|/|\lambda_{11}| = 1.73 \times 10^{-5}$. Our desired expected value of the surface roughness is 6.53.

Using Eq. (5.87), we design a nonlinear state feedback controller such that $\lambda_{c\alpha_i} = \lambda_{c\beta_i} = -0.0373$, for $i = 1, \ldots, 10$. Closed-loop simulations are performed to study the evolution of the expected value of the surface roughness under nonlinear state feedback control. Closed-loop surface roughness profiles obtained from 100 independent simulation runs using the same simulation parameters are averaged; the resulting closed-loop expected surface roughness profile is shown in Fig. 5.13 (solid line), where it is compared with the open-loop expected surface roughness profile (dotted line). We can see that the controller successfully drives the surface roughness to the desired level, which is lower than that corresponding to the open-loop operation $[u_i(t) = 0, i = 1, \ldots, 20]$.

Comparison of Closed-Loop Performance Under Nonlinear and Linear Control

In this subsection, we demonstrate that the performance of the proposed nonlinear controller is superior to that of the linear state feedback controller resulting from the linearization of the nonlinear controller around the zero solution. The nonlinear controller is the same as that presented in Section 5.4.4. The linear state feedback controller design is based on a 20th-order stochastic ODE approximation constructed using the first 20 eigenmodes of the system in Eq. (5.70) as follows:

$$u = B_s^{-1} \left(\Lambda_{cs} - \Lambda_s \right) \tilde{x}_s,$$
(5.112)

where the definitions of the matrices B_s, Λ_{cs}, and Λ_s are the same as those in Eq. (5.79) and $\tilde{x}_s = [\alpha_1 \cdots \alpha_{10} \beta_1 \cdots \beta_{10}]^T$. Twenty control actuators are used to control the system. The ith actuator distribution function is the same as that shown in Eq. (5.111). The desired expected value of the surface roughness is 6.53, and the linear state feedback controller is designed such that $\lambda_{c\alpha_i} = \lambda_{c\beta_i} = -0.0373$, for $i = 1, \ldots, 10$. The linear controller is also applied to the 200th-order approximation of the nonlinear stochastic KSE.

Closed-loop simulations are carried out to evaluate the performance of the closed-loop system achieved under the nonlinear controller and to compare it to that of the corresponding linear controller. Two cases are studied. In both cases, the desired expected surface roughness is 6.53 and the initial surface roughness is chosen to be 18 and 45, respectively. In each case, closed-loop simulation runs are carried out using both the linear state feedback controller [Eq. (5.112)] and the nonlinear state feedback controller [Eq. (5.79)]. Further, in each case, the closed-loop surface roughness profiles are obtained by averaging 100 independent closed-loop simulation runs using the same simulation parameters; the resulting expected surface roughness profiles are presented in Fig. 5.14. Moreover, in both cases, closed-loop surface roughness profiles under the linear and the nonlinear state feedback controllers obtained from a single simulation run using the same simulation parameters are presented in Fig. 5.15.

The number of actuators used by both the linear controller and the nonlinear controller is the same (20 control actuators for all the closed-loop simulations discussed in this subsection). The parameters for the two controllers are the same except that the linear controller excludes the nonlinear term of the stochastic KSE. As a result, when the initial condition is small, both controllers are able to drive the expected closed-loop surface roughness to the desired level. However, based on the simulation results presented in Fig. 5.14, it is clear that a large initial condition deteriorates the closed-loop performance under the linear controller while the nonlinear controller consistently achieves good closed-loop performance independently of the initial condition. This is because under the nonlinear feedback controller, the dominant modes of the closed-loop system in Eq. (5.81) (in particular, the first 20 modes) can be approximated by a stable linear stochastic system [see also Eqs. (5.88) and (5.98) in the Appendix]. In a stable linear stochastic system, the decay rate of the expected value of the surface roughness to the set-point value depends on the eigenvalues of the matrix Λ_{cs} (assigned by the nonlinear controller) and is independent of the initial value of the state (see, for example, [101, 115, 117]). Therefore, the closed-loop performance (in terms of the rate of convergence of the expected value of surface roughness to the set point) under the nonlinear controller is practically independent of the initial condition.

Remark 5.9. Note that a surface roughness profile obtained from one simulation run is one realization of a stochastic process. Due to the stochastic nature of the process, surface roughness profiles from different simulation runs using the same simulation parameters are not identical, but will be around the expected surface roughness. Also, stochastic fluctuations can be observed in all simulation results. By averaging the

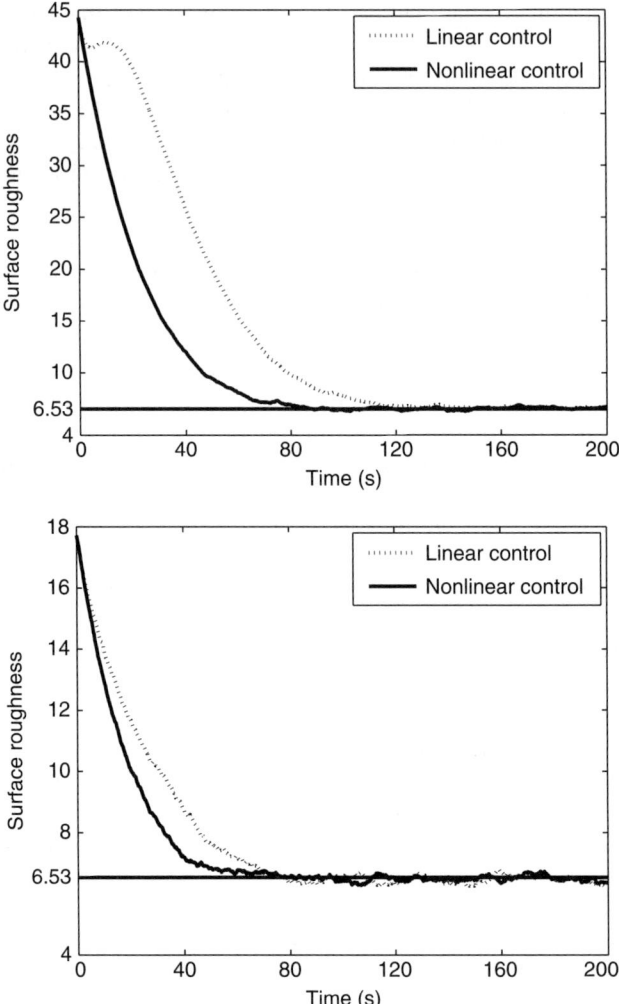

Fig. 5.14. Comparison of the expected closed-loop surface roughness under the nonlinear controller (solid line) and that of the linear controller (dotted line) when the initial surface roughness is 18 (top) and 45 (bottom). The nonlinear controller's performance is superior to that of the linear controller.

surface roughness profiles from multiple independent simulation runs, the stochastic fluctuation can be reduced and the resulting profile will be closer to the expected surface roughness. Conceptually, if we run a very large number of simulations with the same parameters, and average the roughness profiles obtained from each simulation run, the desired expected roughness profile can be obtained. In this study, we compute the expected values of surface roughness in both the open-loop and closed-loop simulations by averaging surface roughness profiles obtained from 100 independent

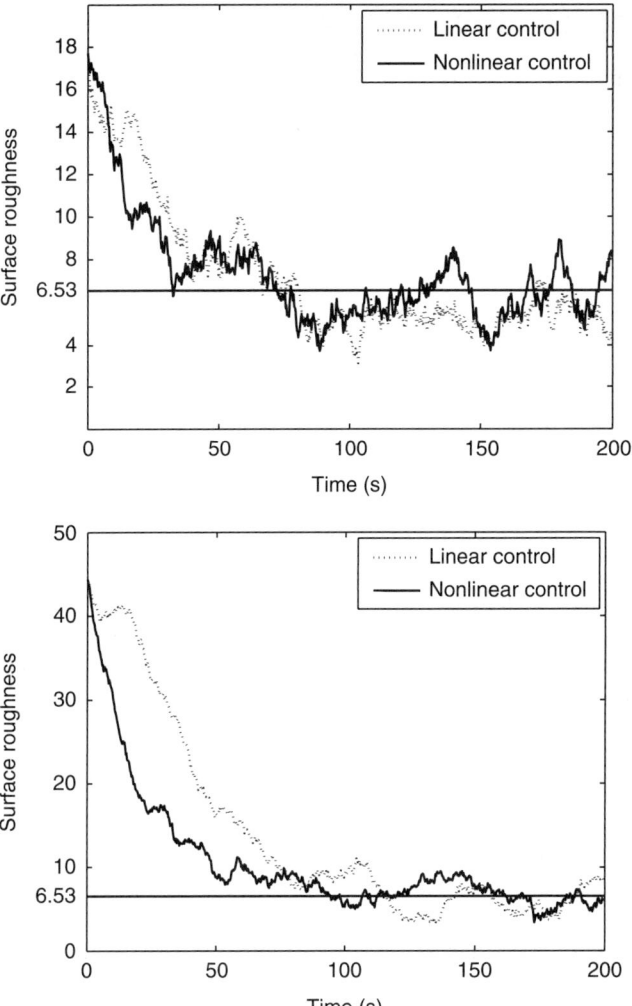

Fig. 5.15. Comparison of the closed-loop surface roughness under the nonlinear controller (solid line) and that of the linear controller (dotted line) from a single simulation run when the initial surface roughness is 18 (top) and 45 (bottom). The nonlinear controller's performance is superior to that of the linear controller.

simulation runs. The resulting surface roughness profiles have few stochastic fluctuations and are very close to the expected values. Furthermore, our control objective is to control the expected surface roughness to a desired level. In practice, a lower surface roughness is usually preferred. In our control problem formulation, if the expected surface roughness is controlled to a lower level compared to that in the open-loop operation, it is expected that the surface roughness from each run will be lowered.

5.4.5 Application to a KMC Model of an Ion-Sputtering Process

In this section, we demonstrate nonlinear control of the sputtering process described in Section 4.4.5, which includes two surface microprocesses, diffusion and erosion. The proposed nonlinear feedback controller is applied to the kinetic Monte Carlo process model of the sputtering process to control the surface roughness to a desired level. The equation for the height fluctuations of the surface in this sputtering process was derived in [91] and is a stochastic Kuramoto–Sivashinsky equation of the form

$$\frac{\partial h}{\partial t} = -\nu \frac{\partial^2 h}{\partial x^2} - \kappa \frac{\partial^4 h}{\partial x^4} + \frac{\lambda}{2} \left(\frac{\partial h}{\partial x} \right)^2 + \xi(x, t), \qquad (5.113)$$

where $x \in [-\pi, \pi]$ is the spatial coordinate, t is the time, $h(x,t)$ is the height of the surface at position x and time t, ν and κ are two constants, and $\xi(x,t)$ is a Gaussian noise with zero mean and covariance:

$$\langle \xi(x,t)\xi(x',t') \rangle = \sigma^2 \delta(x - x')\delta(t - t'), \qquad (5.114)$$

where σ is a constant, $\delta(\cdot)$ is the Dirac function, and $\langle \cdot \rangle$ denotes the expected value. Note that the noise covariance depends on both space x and time t. We note that this stochastic KSE representation for the surface morphological evolution in sputtering processes is limited to surface morphologies that do not involve re-entrant features; the re-entrant features could arise under certain sputtering conditions and are catastrophic for the surface.

Open-Loop Dynamics of the Sputtering Process

In this section, we compute the expected surface roughness profile of the sputtering process by using both the kinetic Monte Carlo model and the stochastic KSE model of the process. The kMC simulation algorithm described in Section 4.4.5 is used. Upon the execution of one Monte Carlo event, α_n or β_n are updated. If the executed event is erosion, α_n or β_n can be updated using Eq. (5.115). If the executed event is diffusion from site i to site j, α_n or β_n can be updated by using Eq. (5.116):

$$\alpha_n^{\text{new}} = \alpha_n^{\text{old}} + \frac{a\left[\psi(n, z_i - a/2) - \psi(n, z_i + a/2)\right]}{n},$$

$$\beta_n^{\text{new}} = \beta_n^{\text{old}} + \frac{a\left[\phi(n, z_i + a/2) - \phi(n, z_i - a/2)\right]}{n}, \qquad (5.115)$$

$$\alpha_n^{\text{new}} = \alpha_n^{\text{old}} + a \cdot \left\{ \frac{\left[\psi(n, z_i - a/2) - \psi(n, z_i + a/2)\right]}{n} \right.$$
$$\left. - \frac{\left[\psi(n, z_j - a/2) - \psi(n, z_j + a/2)\right]}{n} \right\}, \qquad (5.116)$$

$$\beta_n^{\text{new}} = \beta_n^{\text{old}} + a \cdot \left\{ \frac{\left[\phi(n, z_i + a/2) - \phi(n, z_i - a/2)\right]}{n} \right.$$
$$\left. - \frac{\left[\phi(n, z_j + a/2) - \phi(n, z_j - a/2)\right]}{n} \right\},$$

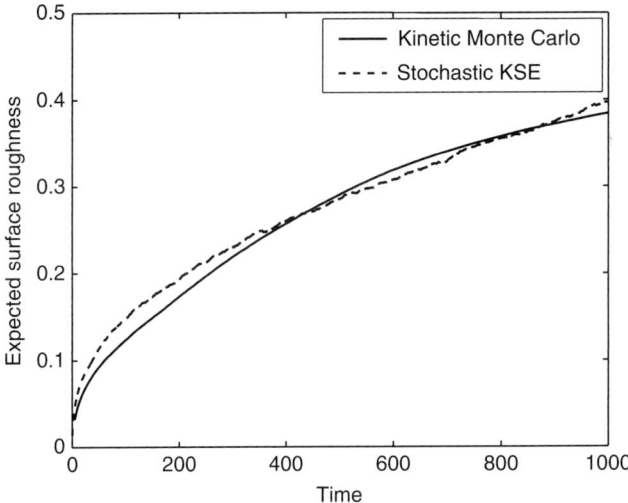

Fig. 5.16. Comparison of the open-loop profile of the expected surface roughness from the kinetic Monte Carlo simulation (solid line) and that from the solution of the stochastic KSE process model (dotted line).

where a is the lattice parameter and z_i is the coordinate of the center of site i.

We also compute the expected surface roughness of the sputtering process based on its stochastic KSE process model of Eq. (5.113). A 200th-order stochastic ordinary differential equation approximation of the system in Eq. (5.113) obtained via Galerkin's method is used to simulate the process (the use of higher-order approximations led to identical numerical results, thereby implying that the following simulation runs are independent of the discretization). The δ function involved in the covariances of ξ_α^n and ξ_β^n is approximated by $1/\Delta t$, where Δt is the integration time step. The parameters of the stochastic KSE model are $\nu = 3.27 \times 10^{-6}$, $\kappa = 1.34 \times 10^{-8}$, $\lambda = 7.52 \times 10^{-6}$, and $\sigma = 4.65 \times 10^{-3}$.

In Fig. 5.16, we compare the expected value of the open-loop surface roughness of the sputtering process from the solution of the stochastic KSE model in Eq. (5.113) to that from a kinetic Monte Carlo simulation. The two profiles are very close. Therefore, by using the stochastic KSE model in Eq. (5.113), we can predict the evolution of the expected surface roughness in this sputtering process. This stochastic KSE model is used as the basis for feedback controller design.

Feedback Control Design

Our control objective is to control the expected surface roughness in the sputtering process to a desired value. Based on the stochastic KSE model of the sputtering process [Eq. (5.113)], a distributed control problem is formulated by following Eq. (5.108). We design a state feedback controller based on a 40th-order stochastic

ODE approximation constructed by using the first 40 eigenmodes of the stochastic KSE model in Eq. (5.113). Forty control actuators are used to control the system. The ith actuator distribution function is taken to be

$$
b_i(z) = \begin{cases} \dfrac{1}{\sqrt{\pi}} \sin(iz), & i = 1, \ldots, 20, \\[2ex] \dfrac{1}{\sqrt{\pi}} \cos[(i-20)z], & i = 21, \ldots, 40. \end{cases} \tag{5.117}
$$

The desired closed-loop surface roughness is 0.3 in this simulation. We design the state feedback controller such that $\lambda_{c\alpha_i} = \lambda_{c\beta_i} = -0.01$, for $i = 1, \ldots, 20$.

In the sputtering process described in Section 4.4.5, the probability f in Eq. (4.92) is dependent on the operating conditions of the sputtering process. Based on the process description, the value of f affects the ratio of erosion and diffusion events on the surface. Since an erosion event is a direct consequence of the bombardment by incoming particles, a higher bombardment rate will result in a higher erosion rate, which implies a larger f. On the other hand, the surface diffusion rate, r_d in Eq. (4.92), should not depend on the bombardment rate of incoming particles. When spatially distributed control is implemented, the surface bombardment rate is a spatially distributed variable. Consequently, f is a spatially distributed variable that can be computed based on the surface bombardment rate.

The controller can be implemented by manipulating the probability that a randomly selected site is subject to the erosion rule, f. From a practical point of view, a spatially distributed erosion probability can be realized by varying the bombardment rate across the substrate. Specifically, the bombardment rate of each surface site under feedback control is $1/\tau = 1 + (\sum_{j=1}^{2m} b_j(z_i)u_j(t))/a$. Since the variation of the bombardment rate does not change the surface diffusion rate, according to the discussion above, the f of site i should relate to the surface bombardment rate in a way that $(1-f)/\tau$ is a constant. Since, in open-loop operation, $\bar{f} = 0.5$ and $1/\bar{\tau} = 1$, we have $(1-f)/\tau = (1-\bar{f})/\bar{\tau} = 0.5$. Therefore, f under feedback control is determined according to the following expression:

$$
f(i) = \frac{\bar{f} + \left(\displaystyle\sum_{j=1}^{2m} b_j(z_i)u_j(t) \right)/a}{1 + \left(\displaystyle\sum_{j=1}^{2m} b_j(z_i)u_j(t) \right)/a}, \tag{5.118}
$$

where $\bar{f} = 0.5$ is the probability that a selected surface site is subject to erosion, and $1/\bar{\tau} = 1$ is the bombardment rate of each surface site in an open-loop operation.

Finally, we apply the designed controller to the kinetic Monte Carlo model of the sputtering process to control the surface roughness to the desired level. In this simulation, the initial surface roughness is about 0.5 and the microstructure of the initial surface is shown in Fig. 5.17.

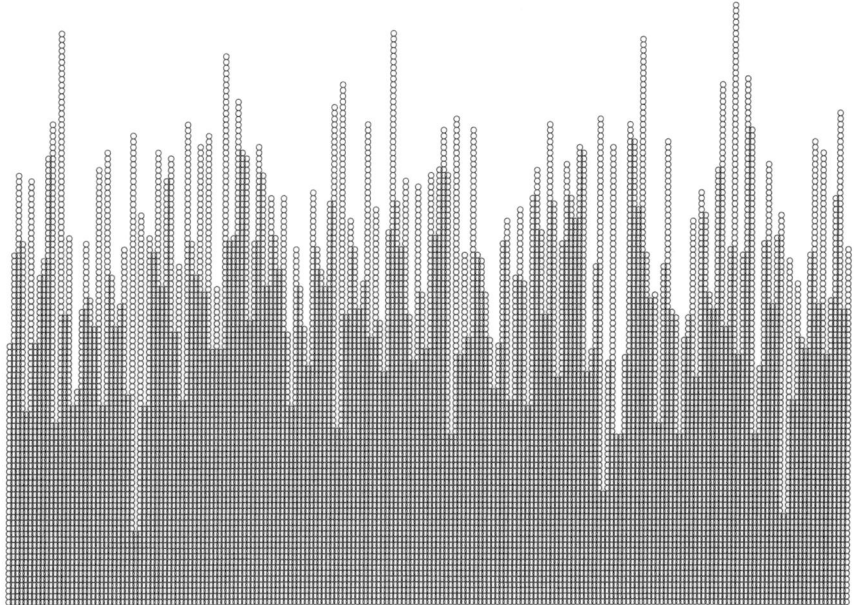

Fig. 5.17. Surface microconfiguration at the beginning of the closed-loop simulation run.

The following simulation algorithm is used to run the kinetic Monte Carlo simulations for the closed-loop system. First, a random number ζ_1 is generated to pick a site i among all the sites on the 1D lattice; the probability that a surface site is subject to the erosion rules, $f(i)$, is determined using Eq. (5.118). Then, the second random number, ζ_2, distributed uniformly in the $(0, 1)$ interval is generated. If $\zeta_2 < f(i)$, site i is subject to erosion; otherwise, the site is subject to diffusion.

If site i is subject to erosion, P_e is computed using the box rule shown in Fig. 4.12 with the box centering the surface particle on site i and $Y(\phi_i)$ is computed using Eq. (4.88). Then, another random number, ζ_{e3}, distributed uniformly in the $(0, 1)$ interval is generated. If $\zeta_{e3} < P_e \cdot Y(\phi_i)$, the surface particle on site i is removed. Otherwise, no Monte Carlo event is executed.

If site i is subject to diffusion, a side neighbor, $j = i + 1$ or $i - 1$, is randomly picked with equal probability and the probability of a hopping from site i to site j, $w_{i \rightarrow j}$, is computed based on Eq. (4.90). Then, another random number, ζ_{d3}, in the $(0, 1)$ interval, is generated. If $\zeta_{e3} < w_{i \rightarrow j}$, the surface particle on site i is moved to site j. Otherwise, no Monte Carlo event is executed. Once a Monte Carlo event is executed, the first 40 states ($\alpha_1, \ldots, \alpha_{20}$ and $\beta_1, \ldots, \beta_{20}$) are updated and new control actions are computed to update the value of f [defined in Eq. (5.118)] for each surface site.

The closed-loop system simulation result is shown in Fig. 5.18. The expected surface roughness, which is the average of the surface roughness profiles obtained from 100 independent runs, under feedback control is plotted as a solid line. We can see that the controller successfully drives the expected surface roughness to the

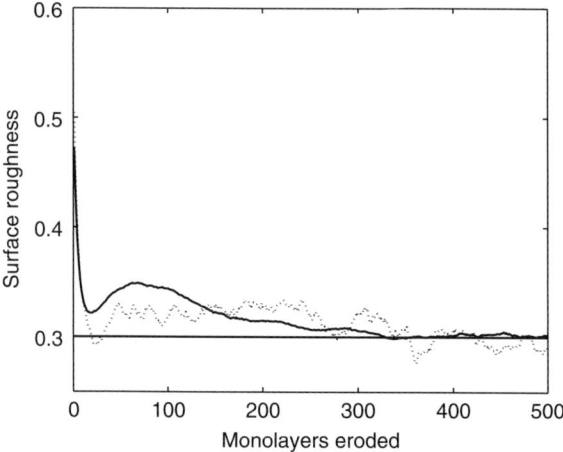

Fig. 5.18. Closed-loop surface roughness profiles in the sputtering process: (a) The expected closed-loop surface roughness profile obtained from 100 independent simulation runs (solid line); (b) the closed-loop surface roughness profile from one simulation run (dotted line).

desired value. The dotted line shows the surface roughness profile under feedback control from one simulation run; due to the stochastic nature of the sputtering process, stochastic fluctuations can be observed in the closed-loop surface roughness profile, but the surface roughness is very close to the expected surface roughness under nonlinear feedback control. We can see that under nonlinear feedback control, the surface roughness can be controlled to the desired level.

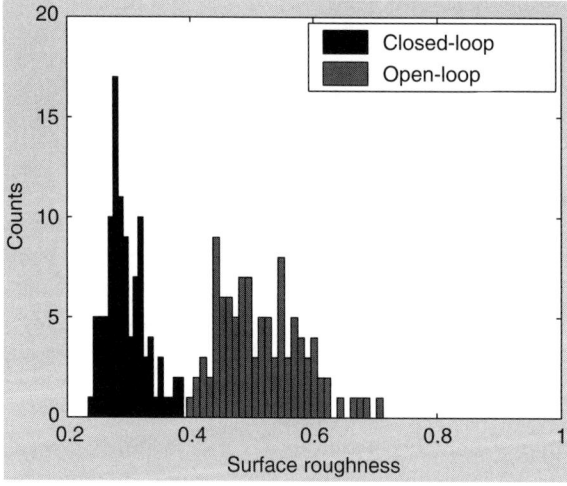

Fig. 5.19. Histogram of the final surface roughness of 100 closed-loop and 100 open-loop simulation runs.

Fig. 5.20. Surface microconfiguration at the end of the closed-loop simulation run under nonlinear feedback control.

Figure 5.19 shows the final surface roughness histogram after 500 monolayers are eroded using 100 different closed-loop simulation runs and that using 100 different open-loop simulation runs. It is clear that the surface roughness from closed-loop simulation runs is lower than that from open-loop simulations runs. Moreover, the variance of the final surface roughness from 100 closed-loop simulation runs is 0.067%, while the variance of the final surface roughness from 100 open-loop simulation runs is 0.53%. The relatively larger variance among the final surface roughnesses by open-loop simulations can be attributed to the stochastic nature of the sputtering process itself. As demonstrated in Fig. 5.19, feedback control can not only reduce the expected final surface roughness, but can also effectively reduce the variance of the final surface roughness.

The microstructure of the surface at the end of the closed-loop system simulation run is shown in Fig. 5.20. It is clear that the proposed nonlinear feedback control method can reduce the surface roughness to the desired level.

5.5 Conclusions

This chapter developed methods for model-based controller design based on stochastic PDEs to control the thin-film surface roughness. A method for multivariable model-predictive control using linear stochastic PDEs was first presented. The method resulted in the design of a computationally efficient multivariable predictive control algorithm, which was successfully applied to the kMC models of thin-film deposition processes taking place on both 1D and 2D lattices to regulate the thin-film thickness and surface roughness at desired levels. Furthermore, it was found that the surface roughness can be regulated using the covariance control technique when spatially distributed sensing and actuation are available. Methods for linear and nonlinear covariance controller design were also developed and were applied to the kMC models of a deposition process and of an ion-sputtering process to control the surface roughness to desired levels.

6

Optimization of Multiscale Process Systems

6.1 Introduction

Optimal operation of processes with respect to certain economic criteria has always been an industrial priority and a subject of intense research. The problem is gaining considerable significance due to increased market competition, reduction in profit margins, and increasingly stringent environmental constraints. This has led to the incorporation of detailed and sophisticated process models into process optimization frameworks.

Multiscale systems are characterized by both macroscopic and microscopic phenomena. Macroscopic process models for such systems involve time-dependent partial differential equations (PDEs), whose spatial discretization results in a set of equality constraints in the form of a set of ordinary differential equations (ODEs). Optimization, then, may involve further discretization of the ODEs in time to derive algebraic equalities and yield a large nonlinear program (NLP) [9, 20] or alternatively discretization in time of only the design variables and the direct integration of the ODEs, keeping track of the inequality constraints, thus formulating a comparably smaller NLP (a scheme known as control vector parameterization (CVP) [153, 41, 21, 19, 131]; semi-infinite programming approaches can also be considered [122, 69]). Optimization methodologies have also been developed when closed-form process descriptions are unavailable or black-box simulations form the equality constraints [113, 84]. Recently, algorithms to efficiently handle noisy objective functions or objective functions that display multiscale behavior have also been developed. These include funneling algorithms, derivative-free optimization algorithms, etc. [2, 22, 104]. However, efficient optimization problem formulations for multiscale process models defined over disparate length scales are lacking.

Motivated by the above considerations, the present chapter and the next chapter present methods for the optimization of multiscale process systems. Specifically, the present chapter addresses the issue of efficient solution of optimization problems when the cost functional and/or equality constraints span multiple length scales and necessitate multiscale process models. We consider a conceptual thin-film epitaxy

P.D. Christofides et al., *Control and Optimization of Multiscale Process Systems*,
Control Engineering, DOI 10.1007/978-0-8176-4793-3_6,
© Birkhäuser Boston, a part of Springer Science+Business Media, LLC 2009

process and optimize the process operation for two simultaneous objectives that span multiple length scales: (1) to maximize the thickness uniformity of the deposited film (macroscopic objective) and (2) to minimize the surface roughness of the deposited film (microscopic objective) across the wafer surface at the end of the process cycle. A multiscale process model is formulated linking steady-state continuum conservation laws for reactor-scale phenomena and kinetic Monte Carlo (kMC) simulations for the microscopic film-surface processes. The computational intensity of the multiscale process model prohibits its direct incorporation into process optimization. To address this issue, order-reduction techniques for dissipative PDEs are linked with the adaptive tabulation scheme for the solution data from the microscopic model to derive a computationally efficient multiscale model that forms the equality constraints of the optimization problem. Initially, we calculate optimal substrate temperature profiles for a steady-state process operation such that the grown thin films have a high degree of spatial uniformity and, simultaneously, low surface roughness. Subsequently, we improve the process operation by computing time-varying substrate temperature radial profiles and inlet concentration profiles of the precursors to meet the optimization objectives. The results of this chapter were first presented in [150, 151].

6.2 Optimization Problem Formulation

Mathematically, the spatially distributed and multiscale process can be represented as

$$0 = \mathcal{A}(x) + f(x, d), \quad \text{on } \Omega_1,$$

$$d(z, t) = \sum_i^n d_i(z)(H(t - \bar{t}_i) - H(t - \bar{t}_{i+1})), \tag{6.1}$$

$$x_m(t_i) = \Pi(x_m(t_{i-1}), \delta t, x|_\gamma), \quad \text{on } \Omega_2,$$
$$\delta t = t_i - t_{i-1}, \tag{6.2}$$

$$g\left(x, \frac{dx}{d\eta}\right) = 0, \quad \text{on } \Gamma \setminus \gamma, \tag{6.3}$$

$$h\left(\bar{x}_s, x|_\gamma, \frac{dx}{d\eta}\right) = 0, \quad \text{on } \gamma. \tag{6.4}$$

Equations (6.1) and (6.2) represent the macroscopic and microscopic descriptions of the process over the respective domains Ω_1 and Ω_2. It is assumed that Ω_1 and Ω_2 do not overlap and share a common interface γ and that $\Omega = \Omega_1 \cup \Omega_2$ spans the whole process domain. $x(z) \in \mathbb{R}^N$ denotes the vector of macroscopic state variables, $x_m(t_i)$ is the vector of microscopic state variables at time instant t_i, $z = [z_1, z_2, z_3] \in \Omega_1 \subset \mathbb{R}^3$ is the vector of spatial coordinates, and Γ is the boundary of the macroscopic domain Ω_1. $\mathcal{A}(x)$ is a second-order dissipative, possibly nonlinear, spatial differential operator, $f(x, d)$ is a nonlinear vector function that is assumed to be sufficiently smooth with respect to its arguments, $d \in \mathbb{R}^p$ is the vector of design variables, $(\bar{t}_i, \bar{t}_{i+1}]$ is the time period where the design variable attains a value d_i, $H(\cdot)$ is the standard Heaviside function, and n is the number of such time instants (leading to $n - 1$ time periods). $g(x, dx/d\eta)$, defined on the boundary $\Gamma \setminus \gamma$, is a nonlinear vector function that is assumed to be sufficiently smooth (the operator $\setminus$ denotes set subtraction), and η is the spatial direction perpendicular to the boundary Γ. We assume that the time horizon over which all the dynamics of the eigenmodes of Eq. (6.1) relax, t_i^s, is negligible in comparison to $\delta \bar{t}_i = \bar{t}_{i+1} - \bar{t}_i$, implying that quasi-steady-state approximation for the process can be assumed.

Function Π can be thought of as a black-box time-stepper (i.e., integrator of the microscopic model), which interacts via an input–output structure and whose internal details may be unknown. It uses $x_m(t_{i-1})$ and the macroscopic state at the interface γ as input, evolves over the time interval δt, and produces state $x_m(t_i)$. The vector function $h(\bar{x}_s, x|_\gamma, dx/d\eta)$ represents the boundary conditions at the common interface between the macroscopic and microscopic domains, γ (thus linking the macroscopic system with the microscopic system), and $\bar{x}_s$ represents the stationary state of the "coarse" realization (the optimization-oriented metrics of the microscopic state), $\bar{x}$, of x_m. It is assumed that such a stationary state exists and is independent of the initial microscopic state, i.e., $x_m(t = 0)$. Coarse variables, $\bar{x}$, are defined through the following restriction operator:

$$\bar{x} = L(x_m). \tag{6.5}$$

The inverse of restriction, termed "lifting", is defined by the following nonunique operator:

$$x_m = l(\bar{x}). \tag{6.6}$$

These two operations provide the bidirectional communication across the different length scales. Typically, when the dynamic behavior of the coarse variables is of interest, one starts with a set of initial coarse variables, generates a number of consistent microscopic initial realizations through lifting, evolves the microscopic system until the desired time horizon, and then restricts the microscopic state to a new set of coarse states. The above operation is usually referred to as a "lift-evolve-restrict procedure". By repeated application of the above scheme, we obtain the dynamic evolution of coarse variables. For a detailed analysis of the lift-evolve-restrict procedure, the reader may refer to [85, 55]. It should be noted that lifting is a one-to-many operation because of a large number of degrees of freedom available in the microscopic configurational space. The set of coarse variables, which form the basis of lifting, must be carefully chosen such that the (unknown) coarse dynamics are observable. Typically, they are lower-order statistical moments of microscopic states; however, some applications may require additional coarse variables derived from spectral moments. It is usually assumed that the higher-order statistical and spectral moments are slaved by these coarse variables. However, in some cases, a "healing" period may be required for the higher-order dynamics to relax [105, 85].

A general optimization problem for the multiscale system in Eqs. (6.1)–(6.4) can be formulated as

$$\min \; \mathcal{G}(x, \bar{x}_s, d, \delta \bar{t}_i) = \sum_i^n \int_\Omega G(x, \bar{x}_s, d, \delta \bar{t}_i) dz$$

$$s.t.$$

$$\mathcal{A}(x) + f(x, d) = 0, \quad \text{on } \Omega_1,$$

$$g\left(x, \frac{dx}{d\eta}\right) = 0 \;\; \text{on } \Gamma \setminus \gamma, \quad h\left(\bar{x}_s, x, \frac{dx}{d\eta}\right) = 0, \quad \text{on } \gamma, \tag{6.7}$$

$$p(x, \bar{x}, d) \leq 0, \quad \forall \, z \in \Omega_1,$$

$$d(z, t) = \sum_i^n d_i(z)(H(t - \bar{t}_i) - H(t - \bar{t}_{i+1})),$$

where $\mathcal{G}(x, \bar{x}_s, d, \delta \bar{t}_i)$ is the objective functional and measures the process performance at both macroscopic and microscopic levels and $p(x, \bar{x}, d)$ is the vector of inequality constraints and may include bounds on the state and design variables; the time intervals $\delta \bar{t}_i$ and design variables d_i are the optimization variables. The microscopic description in the above optimization problem enters implicitly through the parameters $\bar{x}_s$, which is the restriction of the microscopic scale properties in the macroscopic scale. Note that the time-constant process description is a special case of the above when $n = 1$.

Remark 6.1. The condition imposed on the domains of definition of the two process descriptions can be extended to include a finite volume overlap between the two. This overlapping region is used to "reconcile" the two different simulation results and dampen the noise originating from the microscopic solver. In this case, the averaged physical quantities over the overlapping domain, which are computed using the microscopic solver, are required to be consistent with the macroscopic quantities.

6.3 Order Reduction of Dissipative PDEs

We focus on spatially distributed processes modeled by highly dissipative PDE systems with the following state-space description:

$$\frac{\partial x}{\partial t} = \mathcal{A}(x) + f(t, x, d), \quad x(z, 0) = x_0(z),$$

$$q\left(x, \frac{dx}{d\eta}, \ldots, \frac{d^{n_o - 1} x}{d\eta^{n_o - 1}}\right) = 0, \quad \text{on } \Gamma, \tag{6.8}$$

where $x(z, t) \in \mathbb{R}^n$ denotes the vector of state variables, $t \in [0, t_f]$ is the time (t_f is the terminal time), $z = [z_1, z_2, z_3] \in \Omega \subset \mathbb{R}^3$ is the vector of spatial coordinates, Ω is the domain of definition of the process, and Γ is its boundary. $\mathcal{A}(x)$ is a dissipative, possibly nonlinear, spatial differential operator that includes higher-order spatial derivatives, $f(t, x, d)$ is a nonlinear, possibly time-varying, vector function that is assumed to be sufficiently smooth with respect to its arguments, $d(t) \in \mathbb{R}^p$ is the vector of design variables that are assumed to be piecewise-continuous functions of time, $q(x, dx/d\eta, \ldots, d^{n_o-1}x/d\eta^{n_o-1})$ is a nonlinear vector function that is assumed to be sufficiently smooth [n_o, an even number, is the order of the PDE of Eq. (6.8)], $\dfrac{dx}{d\eta}\Big|_\Gamma$ denotes the derivative in the direction perpendicular to the boundary, and $x_0(z)$ is a smooth vector function of z.

The system in Eq. (6.8) arises in the modeling of a wide range of dynamic spatially distributed processes including both transport-reaction processes and several classes of dissipative fluid dynamic systems [28]. The nonlinear structure of the spatial differential operator, $\mathcal{A}(x)$, allows us to account for the explicit dependence of diffusivity and thermal conductivity on temperature and concentration in certain transport-reaction processes, while the nonlinear term $f(t, x, d)$ allows us to model complex reaction mechanisms.

A general optimization problem for the system in Eq. (6.8) can be formulated as follows:

$$\min \int_0^{t_f} \int_\Omega G(x(z, t), d(t)) dz dt$$

s.t.

$$-\frac{\partial x}{\partial t} + \mathcal{A}(x) + f(t, x, d) = 0, \tag{6.9}$$

$$x(z, 0) = x_0(z), \quad q\left(x, \frac{dx}{d\eta}, \ldots, \frac{d^{n_o-1}x}{d\eta^{n_o-1}}\right) = 0 \ \text{ on } \Gamma,$$

$$g(x, d) \leq 0, \quad \forall z \in \Omega, \ t \in [0, t_f],$$

where $\int_0^{t_f} \int_\Omega G(x, d) dz dt$ is the objective functional and $g(x, d)$ is the vector of inequality constraints and may include bounds on the state and design variables. Both $G(x, d)$ and $g(x, d)$ are assumed to be sufficiently smooth functions of their arguments.

It is desired to obtain a finite-dimensional approximation of the infinite-dimensional program developed above through spatial discretization of the imposed PDE constraints. The reader may refer to [122, 69] for alternative methods for semi-infinite programming. To formulate lower-dimensional NLPs, we employ the method of weighted residuals with empirical eigenfunctions as basis functions [9, 17, 18] instead of standard finite-difference-or finite elements-based spatial discretization. The rationale behind this approach is that solutions of highly dissipative PDEs are dominated by a finite (typically small) number of degrees of freedom [145]. For the case of parabolic PDEs with linear differential operators, these can be identified as finite-dimensional slow eigenmodes [28] and can usually be calculated analytically; however, for nonlinear differential operators with spatially varying coefficients, such an analytical solution is, in general, not feasible. Karhunen–Loève expansion (KLE), coupled with the method of snapshots, is an attractive alternative in such situations. In the following, we briefly review the method of weighted residuals, which is followed by a brief description of the KLE.

6.3.1 Method of Weighted Residuals

We derive finite-dimensional approximations of the infinite-dimensional nonlinear program in Eq. (6.9) by using the method of weighted residuals. To simplify the notation, we consider the optimization program in Eq. (6.9) with $n = 1$. In principle, $x(z, t)$ can be represented as an infinite series in terms of a complete set of basis functions $\phi_k(z)$. We can obtain an approximation $x_N(z, t)$ by truncating the series expansion of $x(z, t)$ up to order N, as follows:

$$x_N(z, t) = \sum_{k=1}^{N} a_{kN}(t)\phi_k(z) \xrightarrow{N \to \infty} x(z, t) = \sum_{k=1}^{\infty} a_k(t)\phi_k(z), \qquad (6.10)$$

where $a_{kN}(t)$, $a_k(t)$ are time-varying coefficients.

Substituting the expansion in Eq. (6.10) into Eq. (6.9), multiplying the PDE and the inequality constraints with the weighting functions, $\psi_\nu(z)$, and integrating over the entire spatial domain, the following finite-dimensional dynamic nonlinear program with ODE equality constraints is obtained, where the optimization parameters are the design variables $d(t)$ and the time-varying coefficients $a_{kN}(t)$:

$$\min \int_0^{t_f} \int_\Omega G\left(\sum_{k=1}^{N} a_{kN}(t)\phi_k(z), d\right) dzdt$$

s.t.

$$-\sum_{k=1}^{N} \dot{a}_{kN}\left(\int_\Omega \psi_\nu(z)\phi_k(z)dz\right)$$

$$+ \int_\Omega \psi_\nu(z)\mathcal{A}\left(\sum_{k=1}^{N} a_{kN}(t)\phi_k(z)\right) dz \qquad (6.11)$$

$$+ \int_\Omega \psi_\nu(z)f\left(t, \sum_{k=1}^{N} a_{kN}(t)\phi_k(z), d\right) dz = 0$$

$$\int_\Omega \psi_\nu(z)g\left(\sum_{k=1}^{N} a_{kN}\phi_k(z), d\right) dz \leq 0,$$

where $a_{kN}(t)$ is the approximation of $a_k(t)$ obtained by an Nth-order truncation. From Eq. (6.11), it is clear that the form of the algebraic equalities and inequalities depends on the choice of the weighting functions, as well as on N. Due to the smoothness of the functions $G(x, d)$, $\mathcal{A}(x)$, $f(t, x, d)$, $g(x, d)$ and the completeness of the set of basis functions, $\phi_k(z)$, the nonlinear program in Eq. (6.11) is a well-defined approximation of the infinite-dimensional program in Eq. (6.9) in the sense that the optimal solution of the program in Eq. (6.11) converges to the optimal solution of the program in Eq. (6.9) as $N \to \infty$.

6.3.2 Karhunen–Loève Expansion

In this section, we use the solution data of the system in Eq. (6.8) to construct global basis functions using Karhunen–Loève expansion. The motivation for studying this approach is provided by the occurrence of dominant spatial patterns in the solution of several dissipative PDEs, which should be accounted for in the shape of the basis functions. This approach will be useful in the context of systems of dissipative PDEs that involve nonlinear spatial differential operators and spatially varying coefficients that lead to nonsymmetric solution profiles. KLE is a procedure used to compute an optimal set of empirical eigenfunctions from an appropriately constructed set of solutions of the PDE system in Eq. (6.8), obtained from high-order discretizations (e.g., using standard numerical integration software or process data directly). In this work, the ensemble of solutions is constructed by computing the solutions of the PDE system in Eq. (6.8) for different values of $d(t)$ and different initial conditions. Specifically, we construct a representative ensemble using the following procedure (see also [65, 18] for a detailed discussion on ensemble construction):

- First, we create a set of different initial conditions.
- We then discretize the interval in which each design variable d_m $(m = 1, \dots, p)$ is constrained to be into m_{d_m} (not necessarily equispaced) subintervals. The discrete values of d_m are denoted by $d_{m,j}$, $j = 1, \dots, m_{d_m} - 1$.
- We also descritize the time interval into n_{d_m} time subintervals (also not necessarily equispaced).
- Subsequently, we compute a set of time profiles for each of the design variables $d_m(t)$ by assigning values for $d_m(t)$ at different time instants t_j, $d_{m,j}$, and subsequently computing $d_m(t)$ for the entire time interval of process operation using linear interpolation.
- Finally, we compute an ensemble of PDE solution data for all possible combinations of initial conditions and profiles of $d(t)$.

Application of KLE to this ensemble of data provides an orthogonal set of basis functions (known as empirical eigenfunctions) for the representation of the ensemble, as well as a measure of the relative contribution of each basis function to the total energy (mean-square fluctuation) of the ensemble. A truncated series representation of the ensemble data in terms of the dominant basis functions has a smaller mean-square error than a representation by any other basis of the same dimension [71]. This implies that the projection on the subspace spanned by the empirical eigenfunctions will, on average, contain the most energy possible compared to all other linear decompositions, for a given number of modes. Therefore, the KLE yields the most efficient way for computing the basis functions (corresponding to the largest empirical eigenvalues) capturing the dominant patterns of the ensemble.

For simplicity of the presentation, we describe the KLE in the context of the system in Eq. (6.8) with $n = 1$ and assume that there is available a sufficiently large set of solutions of this system for different values of d, $\{\bar{v}_\kappa\}$, consisting of K sampled states, $\bar{v}_\kappa(z)$ (which are typically called "snapshots"). The reader may refer to [48, 71, 138, 139] for a detailed presentation and analysis of the KLE. We define the ensemble average of snapshots as $< \bar{v}_\kappa > := \frac{1}{K} \sum_{\kappa=1}^{K} \bar{v}_\kappa(z)$ (we note that nonuniform sampling of the snapshots and weighted ensemble average can also be considered; see, for example, [65]). Furthermore, the ensemble average of snapshots $< \bar{v}_\kappa >$ is subtracted from the snapshots, i.e.,

$$v_\kappa = \bar{v}_\kappa - < \bar{v}_\kappa >, \tag{6.12}$$

so that only fluctuations are analyzed. It is useful to analyze fluctuations in deviation variables rather than the actual variables because usually fewer functions are required to fit them [138]. The issue is how to obtain the most typical or characteristic structure (in a sense that will become clear below) $\phi(z)$ among these snapshots $\{v_\kappa\}$. Mathematically, this problem can be posed as that of obtaining a function $\phi(z)$ that maximizes the following objective function:

$$\text{Maximize} \quad \frac{< (\phi, v_\kappa)^2 >}{(\phi, \phi)} \tag{6.13}$$
$$\text{s.t. } (\phi, \phi) = 1, \quad \phi \in L^2([\Omega]),$$

which, in other words, implies that the projection of $\bar{v}_k$ on the subspace spanned by $\phi(z)$ captures maximum energy. Here, (x, y) denotes a complex inner product defined as

$$(x, y) = \int_\Omega \bar{x}(z)y(z)dz. \tag{6.14}$$

The constraint $(\phi, \phi) = 1$ is imposed to ensure that the function $\phi(z)$, computed as a solution of the above maximization problem, is unique. An alternative way to express the constrained optimization problem in Eq. (6.13) is to solve for ϕ such that

$$\frac{d\bar{L}(\phi + \delta\psi)}{d\delta}(\delta = 0) = 0, \quad (\phi, \phi) = 1, \tag{6.15}$$

where $\bar{L} = < (\phi, v_\kappa)^2 > - \lambda((\phi, \phi) - 1)$ is the corresponding Lagrangian functional and δ is a real number.

Using the definitions of inner product and ensemble average, $d\bar{L}(\phi + \delta\psi)/d\delta(\delta = 0)$ can be computed from the following expression:

$$\frac{d\bar{L}(\phi + \delta\psi)}{d\delta}(\delta = 0) = \int_\Omega \left(\left\{ \int_\Omega < v_\kappa(z)v_\kappa(\bar{z}) > \phi(z)dz \right\} - \lambda\phi(\bar{z}) \right) \psi(\bar{z})d\bar{z}. \tag{6.16}$$

Since $\psi(\bar{z})$ is an arbitrary function, the necessary conditions for optimality take the form

$$\int_\Omega < v_\kappa(z)v_\kappa(\bar z) > \phi(z)dz = \lambda\phi(\bar z), \quad (\phi,\phi) = 1. \tag{6.17}$$

If we introduce the two-point correlation function:

$$K(z,\bar z) =< v_\kappa(z)v_\kappa(\bar z) >= \frac{1}{K}\sum_{\kappa=1}^{K} v_\kappa(z)v_\kappa(\bar z) \tag{6.18}$$

and the linear operator:

$$R := \int_\Omega K(z,\bar z)d\bar z, \tag{6.19}$$

the optimality condition in Eq. (6.17) reduces to the following eigenvalue-eigen-function problem of the integral operator:

$$R\phi = \lambda\phi \implies \int_\Omega K(z,\bar z)\phi(\bar z)d\bar z = \lambda\phi(z). \tag{6.20}$$

The computation of the solution to the above integral eigenvalue problem is, in general, a very expensive computational task. To circumvent this problem, Sirovich, in 1987, introduced the method of snapshots [138, 139]. The central idea of this technique is to assume that the requisite eigenfunction, $\phi(z)$, can be expressed as a linear combination of the snapshots, i.e.,

$$\phi(z) = \sum_k c_k v_k(z). \tag{6.21}$$

Substituting the above expression for $\phi(z)$ in Eq. (6.20), we obtain the following eigenvalue problem:

$$\int_\Omega \frac{1}{K}\sum_{\kappa=1}^{K} v_\kappa(z)v_\kappa(\bar z)\sum_{k=1}^{K} c_k v_k(\bar z)d\bar z = \lambda\sum_{k=1}^{K} c_k v_k(z). \tag{6.22}$$

Defining

$$B^{\kappa k} := \frac{1}{K}\int_\Omega v_\kappa(\bar z)v_k(\bar z)d\bar z, \tag{6.23}$$

the eigenvalue problem in Eq. (6.22) can be equivalently written as:

$$Bc = \lambda c. \tag{6.24}$$

The solution to the above eigenvalue problem (which can be obtained by utilizing standard methods from linear algebra) yields the eigenvectors $c = [c_1 \cdots c_K]$, which can be used in Eq. (6.21) to construct the eigenfunction $\phi(z)$. From the structure of the matrix B, it follows that it is symmetric and positive semidefinite; thus, its eigenvalues, λ_κ, $\kappa = 1,\ldots,K$, are real and nonnegative. The relative magnitude of the eigenvalues represents a measure of the fraction of the "energy" embedded in the ensemble captured by the corresponding eigenfunctions. Furthermore, the resulting eigenfunctions form an orthogonal set, i.e.,

$$\int_\Omega \phi_i(z)\phi_j(z)dz = 0, \quad i \neq j. \tag{6.25}$$

Remark 6.2. The value of m_{d_m} should be determined based on the effect of the design variable d_m on the solution of the system in Eq. (6.8) (if, for example, the effect of the variable d_1 is larger than the effect of the variable d_2, then m_{d_1} should be larger than m_{d_2}).

Remark 6.3. It should be noted that the kernel in Eq. (6.17) is not symmetric for cylindrical or spherical geometries [15]. However, the reformulated problem given by Eq. (6.24) is symmetric irrespective of spatial geometry.

Remark 6.4. The basis that we compute using KLE is specific to the process under investigation and independent of the specific optimization problem we try to solve. Therefore, the same basis can be used to perform computationally efficient optimizations with respect to different objective functionals associated with the same underlying set of partial differential equations.

Remark 6.5. Even though it is expected that the use of more basis functions in the series expansion in Eq. (6.10) would improve the accuracy of the computed approximate model in Eq. (6.11), the use of empirical eigenfunctions corresponding to very small eigenvalues should be avoided because such eigenfunctions are contaminated with significant round-off errors.

Remark 6.6. Iterative methods, such as Krylov subspace methods, can be used to reduce the computational cost associated with the computation of the system eigenvalues and eigenfunctions.

Remark 6.7. In addition to the use of empirical eigenfunctions as basis functions, the concept of approximated inertial manifolds can be used to take advantage of the time-scale separation of the spectrum of the highly dissipative spatial differential operator to construct reduced-order models and nonlinear programs [32, 9].

6.4 Optimization of Microscopic Model Using Tabulation

6.4.1 Problem Formulation

We investigate the problem of dynamic optimization for a class of systems described by the following discrete-time description [53]:

$$\mathbf{X}(t_{i+1}) = \Pi(\mathbf{X}(t_i), \delta t_i, \omega; \boldsymbol{\theta}_i),$$
$$\text{for} \quad \delta t_i = t_{i+1} - t_i, \tag{6.26}$$

where $\mathbf{X}(t_{i+1})$ and $\mathbf{X}(t_i) \in \Omega_1 \subset \mathbb{R}^n$ are the vector of states of the system at time instants t_{i+1} and t_i, respectively, $t_{i+1}, t_i \in [0 \ T]$, $\boldsymbol{\theta}_i \in \Omega_1 \subset \mathbb{R}^p$ is the control input vector, which is constant for $t \in (t_i, t_{i+1}]$, and ω is a random walk defined over some measurable space. Most dynamic systems, continuous or discrete, can be expressed in the form given by Eq. (6.26) when only an input–output relationship is required. For example, spatially distributed parabolic partial differential equations

(PDEs), which arise frequently in the modeling transport-reaction processes, assume the above form, where the right-hand side (RHS) is obtained from the appropriate spatial and temporal discretization of the PDE. For systems that are modeled using atomistic simulations, such as kMC, the RHS represents the corresponding evolution rule. The function $\Pi(\cdot)$ in this case is fundamentally different from the one obtained by discretizing PDEs, as it is unavailable in closed form. Microscopic systems, for which the function $\Pi(\cdot)$ is a "black box," are the focus of this section. We assume the following smoothness assumption with respect to parameters for the process $\mathbf{X}(t_i)$:

Assumption 6.1. *The stochastic process $\mathbf{X}(t_i, \boldsymbol{\theta}, \mathbf{x})$ with $\mathbf{X}(0, \boldsymbol{\theta}, \mathbf{x}) = \mathbf{x}$ defined over the probability space $[\Omega, \Sigma, P_{\boldsymbol{\theta}}]$ is twice continuously differentiable with respect to $\boldsymbol{\theta} \in \Theta$ and $\mathbf{x} \in \mathbb{R}^n$ for all $\omega \in \Omega$ with probability 1.*

We are interested in computing an optimal time-varying profile of the control input, $\theta^*(t)$, such that a prespecified objective for the expected process dynamics is realized. Such a profile can be obtained from the solution of the following dynamic optimization problem:

$$\min_{\theta(t)} \int_0^{t_f} \mathcal{Q}(E(\mathbf{X}), \theta)dt + \mathcal{W}(|E(\mathbf{X}(t_f)) - \bar{\mathbf{X}}(t_f)|)$$

$$\text{s.t.} \tag{6.27}$$

$$\mathbf{X}(t_{i+1}) = \Pi(\mathbf{X}(t_i), \delta t_i, \omega; \boldsymbol{\theta}_i), \ \delta t_i = t_{i+1} - t_i, g_d(\mathbf{X}, \theta) \leq 0,$$

where $\mathcal{Q}$ is a scalar cost function, $\mathcal{W}$ is an appropriate final-time penalty function, and g_d denotes the set of inequality constraints on state and manipulated variables. Discretizing the time interval $[0, t_f]$ into N subintervals and assuming $\theta(t)$ to be piecewise constant during each subinterval, we can obtain a finite-dimensional approximation to the above dynamic optimization problem. However, the equality constraints cannot be handled explicitly during optimization due to their unavailability in closed form. The standard approach for the solution of the above optimization problem is to compute the objective functional as a black box during optimization and employ derivative-free optimization algorithms such as Nelder–Mead, Hooke–Jeeves, pattern search [88], etc. to compute $\theta^*(t)$. However, if the computation of the objective functional is expensive, which requires the integration of the microscopic simulator for the period $[0, t_f]$, the solution time required may become prohibitive. To address this issue, we extend the applicability of in situ adaptive tabulation (ISAT) to accomplish efficient simulation of the stochastic time-stepper, resulting in efficient solution of the optimization problem.

6.4.2 In situ Adaptive Tabulation

The in situ adaptive tabulation scheme was originally developed for deterministic systems in the context of efficient implementation of combustion chemistry [123]. This section provides a brief overview of the original algorithm (for details, refer to [123, 150]). Consider a dynamically evolving process with the following state-space description:

$$\dot{\mathbf{x}} = \mathbf{f}(\mathbf{x}, \mathbf{u}) = \mathbf{f}(\boldsymbol{\phi}), \tag{6.28}$$

where $\mathbf{x} \in \Omega_1 \subset \mathbb{R}^n$ is the vector of state variables and $\mathbf{u} \in \Omega_2 \subset \mathbb{R}^p$ is the vector of control variables. The vector $\boldsymbol{\phi}$ is defined as $\boldsymbol{\phi} = [\mathbf{x}\ \mathbf{u}]^T$, $\boldsymbol{\phi} \in \Omega = \Omega_1 \times \Omega_2$. We define $\mathcal{R}(\boldsymbol{\phi}_0)$ to be a nonlinear integral operator representing the evolution of the system from initial state $\boldsymbol{\phi}_0$ at time t_0 to state $\boldsymbol{\phi}(t_0 + \tau) = \mathcal{R}(\boldsymbol{\phi}_0)$ at time $t_0 + \tau$ (the time step τ will henceforth be referred to as the ISAT-reporting horizon). To reduce computational costs, it is desired to approximate $\mathcal{R}(\boldsymbol{\phi}_q)$ due to a "nearby" (in a sense that will become clear later) state $\boldsymbol{\phi}_q$, based on the knowledge of $\{\boldsymbol{\phi}_0, \mathcal{R}\}$. One way to address this issue is to tabulate a large number of doublets $\{\boldsymbol{\phi}, \mathcal{R}(\boldsymbol{\phi})\}$ regularly spanning the whole realizable region Ω into a database [an $(n + p)$-dimensional mesh], and subsequently interpolate within this database to estimate $\mathcal{R}(\boldsymbol{\phi}_q)$. The interpolation error that is incurred can be controlled through refining the mesh. However, the generation of the database, which is usually done in a *preprocessing* phase, can become cumbersome if the dimensionality of the state space (i.e., Ω) is large.

We, define the *accessed* region, Ω_a ($\Omega_a \subset \Omega$), as the set of all states $\boldsymbol{\phi}$ that occur in the calculations. A crucial observation is that the accessed region is much smaller than the realizable region. Exploiting this fact, ISAT constructs the database *online* and hence tabulates only the accessed region, Ω_a. Moreover, to control the interpolation errors, the *mapping gradient matrix* is also computed (and tabulated), which is defined as

$$A_{ij}(\boldsymbol{\phi}) \equiv \frac{\partial \mathcal{R}_i(\boldsymbol{\phi})}{\partial \phi_j}. \tag{6.29}$$

Consider a tabulated triplet $\{\boldsymbol{\phi}_p, \mathcal{R}(\boldsymbol{\phi}_p), \mathbf{A}(\boldsymbol{\phi}_p)\}$. A linear interpolation for $\mathcal{R}(\boldsymbol{\phi}_q)$ can be obtained as

$$\mathcal{R}(\boldsymbol{\phi}_q) \approx \mathcal{R}^l(\boldsymbol{\phi}_q) \equiv \mathcal{R}(\boldsymbol{\phi}_0) + \delta\mathcal{R}^l$$
$$\delta\mathcal{R}^l \equiv \mathbf{A}(\boldsymbol{\phi}_p)\delta\boldsymbol{\phi} + \mathcal{O}(|\,\delta\boldsymbol{\phi}\,|^2), \quad \delta\boldsymbol{\phi} = \boldsymbol{\phi}_q - \boldsymbol{\phi}_p \tag{6.30}$$

The error induced due to the interpolation can be analyzed as follows. Assume that $\boldsymbol{\phi}_p$ and $\boldsymbol{\phi}_q$ are such that $|\,\mathcal{R}(\boldsymbol{\phi}_q) - \mathcal{R}(\boldsymbol{\phi}_p)\,| \leq \epsilon_{\text{tol}}$. It follows from above that

$$\delta\boldsymbol{\phi}^T \mathbf{A}^T(\boldsymbol{\phi}_p)\mathbf{A}(\boldsymbol{\phi}_p)\delta\boldsymbol{\phi} \leq \epsilon_{\text{tol}}^2. \tag{6.31}$$

Equation (6.31) defines a hyper-ellipsoid (referred to as the ellipsoid of attraction, EOA) centered at $\boldsymbol{\phi}_p$ with principle axes given by elements of the diagonal matrix $\boldsymbol{\Sigma}$ such that $\mathbf{Q}^T \boldsymbol{\Sigma} \mathbf{Q}$ is the singular-value decomposition of $\mathbf{A}$. Now given any query $\boldsymbol{\phi}_q$, if there exists a tabulated $\boldsymbol{\phi}_p$ such that Eq. (6.31) is valid, the error due to interpolation will be less than ϵ_{tol}. If such a $\boldsymbol{\phi}_p$ is not found in the database, direct integration of Eq. (6.28) is performed and stored in the database.

The matrix $\mathbf{A}(\boldsymbol{\phi})$ can be related to sensitivity coefficients. The first-order sensitivity coefficients with respect to the initial conditions are defined as

$$B_{ij}(\boldsymbol{\phi}_0, t) \equiv \frac{\partial \phi_i(t)}{\partial \phi_0^j}. \tag{6.32}$$

From the above, it can be seen that

$$\mathbf{A}(\boldsymbol{\phi}_0) = \mathbf{B}(\boldsymbol{\phi}_0, \tau). \tag{6.33}$$

ISAT is implemented in practice using a binary tree. Ideally, once a query point, $\boldsymbol{\phi}_q$, is generated, one would like to determine $\boldsymbol{\phi}_0$ that is *closest* to $\boldsymbol{\phi}_q$ for interpolation purposes. However, a complete database search for $\boldsymbol{\phi}_0$ could be expensive, especially if the database is large. To circumvent this problem, the database is organized as a binary tree comprised of *leafs* and *nodes*. Each node contains the information regarding a convex region that is *likely* to be spanned by the corresponding leafs, which, in turn, contain the *record* comprised of $\boldsymbol{\phi}_0$, $\mathcal{R}(\boldsymbol{\phi}_0)$, $\mathbf{A}(\boldsymbol{\phi}_0)$, Q, and σ. The convex region contained within each node is characterized by a vector $\mathbf{v}$ and scalar a such that leafs pertaining to *subregion* $\mathbf{v}^T \boldsymbol{\phi} < a$ are on the left and leafs corresponding to subregion $\mathbf{v}^T \boldsymbol{\phi} > a$ are on the right. The division into a number of convex regions allows efficient search of the point within the database that is most likely to be nearest the query point.

When the database is probed with a query ($\boldsymbol{\phi}_q$), three distinct possibilities may arise:

1. $\boldsymbol{\phi}_q$ lies within the EOA of $\boldsymbol{\phi}_0$. In this case, presented in Fig. 6.1(a), the corresponding integral map based on interpolation around $\boldsymbol{\phi}_0$ [Eq. (6.30)] is returned.
2. $\boldsymbol{\phi}_q$ lies outside the EOA of $\boldsymbol{\phi}_0$; $\mathcal{R}(\boldsymbol{\phi}_q)$ is computed through simulation, and post-simulation it is observed that $|\mathcal{R}(\boldsymbol{\phi}_q) - \mathcal{R}^l(\boldsymbol{\phi}_q)| < \epsilon_{\text{tol}}$. In this case, presented in Fig. 6.1(b), the EOA around $\boldsymbol{\phi}_0$ is *grown* to include $\boldsymbol{\phi}_q$; the calculated $\mathcal{R}(\boldsymbol{\phi}_q)$ is returned.
3. $\boldsymbol{\phi}_q$ lies outside the EOA of $\boldsymbol{\phi}_0$; $\mathcal{R}(\boldsymbol{\phi}_q)$ is computed through simulation, and post-simulation it is observed that $|\mathcal{R}(\boldsymbol{\phi}_q) - \mathcal{R}^l(\boldsymbol{\phi}_q)| > \epsilon_{\text{tol}}$. In this case, presented in Fig. 6.1(c), the database is augmented by a record for $\boldsymbol{\phi}_q$ and the original leaf, $\boldsymbol{\phi}_0$, is replaced by a node. The records for $\boldsymbol{\phi}_0$ and $\boldsymbol{\phi}_q$ are stored as left and right leafs, respectively, of the new node; the calculated $\mathcal{R}(\boldsymbol{\phi}_q)$ is returned.

In contrast to deterministic black-box systems, the problem of derivative estimation for stochastic black-box systems is complex due to the issues of bias and variance. For example, finite-difference approximations cannot be directly employed in Eq. (6.32) to obtain first-order sensitivity matrix. An extensive amount of literature is available addressing this issue; important techniques include finite difference/finite difference with common random numbers (FD/FDC) [93, 160, 62, 37], infinitesimal perturbation analysis (IPA) [92, 68, 70, 160, 142], and likelihood ratio estimation (LR) [128, 63, 64, 92]. In our implementation of stochastic ISAT, FDC was employed for derivative estimation. In the following subsection we discuss sufficient conditions for unbiasedness and finite variance in derivative estimation using FDC.

6.4.3 FDC Derivative Estimation and EOA

Consider a stochastic process $\mathbf{X}(t, \theta)$, $\mathbf{X} \in \mathbb{R}^n$ $\theta \in \Theta \subset \mathbb{R}^m$, defined over a probability space $(\Omega, \Sigma, P_\theta)$, and let $\mathbf{X}(t, \theta, \omega) \mid t \geq 0$, $\omega \in \Omega$, denote a sample

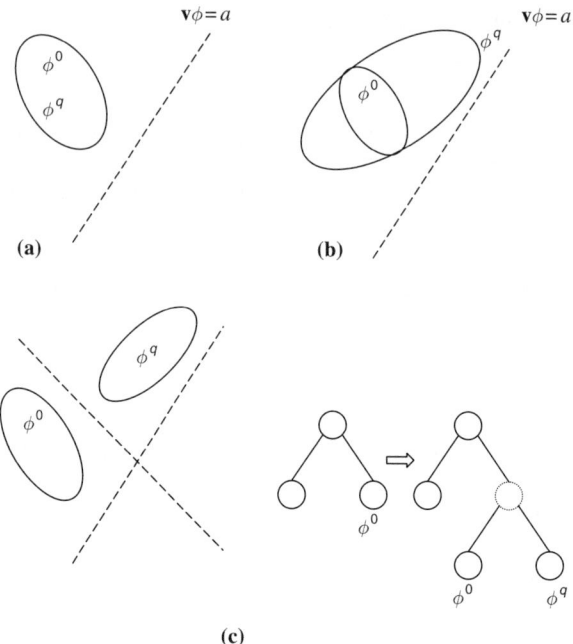

Fig. 6.1. Various possibilities that may arise once the ISAT database is probed with a query. (a) ϕ^q lies within the EOA of ϕ^0. (b) ϕ^q lies outside the EOA of ϕ^0, but $|\mathcal{R}(\phi^q) - \mathcal{R}^l(\phi^q)| < \epsilon_{\text{tol}}$. (c) ϕ^q lies outside the EOA of ϕ^0, and $|\mathcal{R}(\phi^q) - \mathcal{R}^l(\phi^q)| > \epsilon_{\text{tol}}$.

path. For ease of notation, we assume $n = m = 1$ for the rest of the discussion. Let $X'(t, \theta_0)$ be the derivative $\partial X(t, \theta)_{\theta=\theta_0}/\partial\theta$ for some $\theta_0 \in \Theta$, assuming it exists. The FD and FDC estimates of the derivative $X'(t, \theta_0)$ are defined as follows:

$$
\begin{aligned}
\bar{X}'^{,\text{FD}}(t, \theta_0) &= \frac{1}{N}\sum_{i=1}^{N} X_i'^{,\text{FD}}(t, \theta_0), \\
X_i'^{,\text{FD}}(t, \theta_0) &= \frac{X_i(t, \theta_0 + \delta\theta, \omega') - X_i(t, \theta_0, \omega)}{\delta\theta},
\end{aligned}
\tag{6.34}
$$

$$
\begin{aligned}
\bar{X}'^{,\text{FDC}}(t, \theta_0) &= \frac{1}{N}\sum_{i=1}^{N} X_i'^{,\text{FDC}}(t, \theta_0), \\
X_i'^{,\text{FDC}}(t, \theta_0) &= \frac{X_i(t, \theta_0 + \delta\theta, \omega) - X_i(t, \theta_0, \omega)}{\delta\theta}.
\end{aligned}
\tag{6.35}
$$

An immediate issue arising due to the above definitions is the appropriate choice of N and $\delta\theta$ that would guarantee satisfactory unbiasedness and accuracy of the derivative estimates. To make these concepts more precise, we define the following *loss function* [93]:

Definition 6.1. *The loss function associated with a derivative estimator* $\bar{X}'(t, \theta_0)$ *based on N samples is defined as*

$$R_N = E[\bar{X}'(t, \theta_0) - X'(t, \theta_0)]^2 = \text{VAR}(\bar{X}'(t, \theta_0)) + B_N^2,$$
$$B_N = E[\bar{X}'(t, \theta_0)] - X'(t, \theta_0), \qquad (6.36)$$

where the first term denotes the variance of the derivative estimators and the second term denotes the bias. Also, the convergence rate is said to be $\mathcal{O}(f(N))$ *if* $R_N \in \mathcal{O}(f(N))$.

For the variance of the FD and FDC estimators, we state the following result from [62]:

Theorem 6.1. *Suppose that* $X(t, \theta, \omega)$ *is described by Eq. (6.26) and Assumption 6.1 holds. Then, for* $\theta_0 \in \Theta$, $\text{VAR}[X(t, \theta_0 + \delta\theta, \omega') - X(t, \theta_0 + \delta\theta, \omega)]$ *is*

(i) $\mathcal{O}(1)$ *if* ω *and* ω' *are independent.*
(ii) $\mathcal{O}(\delta\theta^2)$ *if* $\omega = \omega'$.

Theorem 6.1 states that variance of FD derivative estimators tends to infinity as $\delta\theta \to 0$. However, using FDC, the variance can be made vanishingly small. Next, we state the following theorem from [93] to establish the convergence of FDC estimates:

Theorem 6.2. *Suppose that Assumption 6.1 holds and* $\Psi(\omega)$ *defined as*

$$\Psi(\omega) = \begin{cases} \sup_{\theta \in \Theta} |\bar{X}'(\theta, \omega)| & \text{if } \omega \in \Omega, \\ 0 & \text{otherwise,} \end{cases}$$

is such that $\Psi(\omega) \le \Gamma(\omega)$ *for some function* $\Gamma : \Omega \to \mathbb{R}$. *If* $\Gamma(\omega)$ *satisfies* $\int_\Omega [\Gamma(\omega)]^2 dP(\omega) < \infty$, *then*

$$R_N^{\text{FDC}} = \sigma_{\text{FDC}}^2/N + [X''(\xi^+)\delta\theta/4]^2, \qquad (6.37)$$

where $\theta_0 \le \xi^+ \le \theta_0 + \delta\theta$. *As a consequence, the convergence rate of FDC estimates is* $\mathcal{O}(N^{-1/2})$ *provided that* $\delta\theta \in \mathcal{O}(N^{-1/2})$. *In the limit* $\delta\theta \to 0$, *the bias also vanishes.*

A combination of Theorems 6.1 and 6.2 forms the theoretical rationale behind the computation of derivatives based on finite differences with common random numbers. From the simulation point of view, FDC can be implemented by resetting the random seed of the random number generator while evaluating $X(t, \theta_0 + \delta\theta, \omega)$ and $X(t, \theta_0, \omega)$.

We now formally define the EOA for systems governed by equations of the form of Eq. (6.26):

Definition 6.2. *Let* $\mathbf{X}(t, \alpha, \mathbf{x})$, $\mathbf{X}(\cdot) \in \mathbb{R}^n$ *be a stochastic process governed by Eq. (6.26) such that* $\mathbf{X}(0, \alpha, \mathbf{x}) = \mathbf{x}$ *and let* $\mathcal{G} : \mathbb{R}^n \times \mathbb{R}^p \to \mathbb{R}^{n \times (n+p)}$ *be the first-order sensitivity matrix defined as*

$$\mathcal{G} = \begin{bmatrix} \dfrac{\partial \mathbf{X}}{\partial \mathbf{x}} & \dfrac{\partial \mathbf{X}}{\partial \alpha} \end{bmatrix}. \tag{6.38}$$

Then the state $\mathbf{z}' = [\mathbf{x}' \ \alpha']^T$, $\mathbf{z} \in \mathbb{R}^{n+p}$, *lies within the ellipsoid of attraction of* $\mathbf{z} = [\mathbf{x} \ \alpha]^T$ *if*

$$(\mathbf{z}' - \mathbf{z})^T \mathcal{G}^T \mathcal{G} (\mathbf{z}' - \mathbf{z}) \leq \epsilon_{\text{tol}}^2. \tag{6.39}$$

In the next section, we present two applications of the above scheme when the underlying dynamical system is modeled by a time-stepper-based description (i.e., microscopic simulator).

6.5 Multiscale Optimization Problem Solution

Finite-dimensional approximations to the infinite-dimensional program in Eq. (6.7) can be obtained through spatial discretization of the PDE equality constraints and of the objective functional to generate a nonlinear program (NLP). Brute-force spatial discretization employing finite differences or finite elements typically results in a large set of algebraic equations, and subsequent storage and computational requirements of the formulated NLP may become prohibitive, requiring the use of specially designed algorithms for large-scale optimization problems. The presence of black-box time-steppers, through the coarse stationary states $(\bar{x}_s)$, into the multiscale model further increases the computational demands. To address this issue, nonlinear order reduction for PDEs using Karhuenen–Loéve expansion (KLE) (Section 6.3) is coupled with in situ adaptive tabulation of $\bar{x}_s$ (Section 6.4) to formulate a reduced-order multiscale model that can be employed to efficiently solve multiscale optimization problems.

6.5.1 Solution Algorithm

A solution algorithm that is applicable to a broad class of multiscale processes modeled by Eqs. (6.1)–(6.4) is outlined below.

1. Formulate a reduced-order macroscopic model employing KLE as described in Section 6.3.
2. Select an arbitrary (but physically consistent) initial condition $x_m(t = 0)$ and $x|_\gamma$, and evolve the black-box time-stepper until $\bar{x}$ reaches a constant value, denoted as $\bar{x}_s$ (the microscopic system rests at stationary state).
3. Solve Eq. (6.1) subject to boundary conditions given by Eqs. (6.3) and (6.4), either analytically or numerically, to obtain new $x|_\gamma$ denoted as x'_i.
4. Use ISAT to evaluate $\bar{x}_s$ based on sampled data. If the error is predicted to be high, simulate the microscopic system until it reaches a new stationary state.
5. Repeat steps 3 and 4 to obtain x'_{i+1}, until the error between x'_i and x'_{i+1} is within an acceptable tolerance.

A flowchart of the above algorithm is presented in Fig. 6.2 for more clarity. The proposed iterative method is used to obtain the solution to the multiscale process

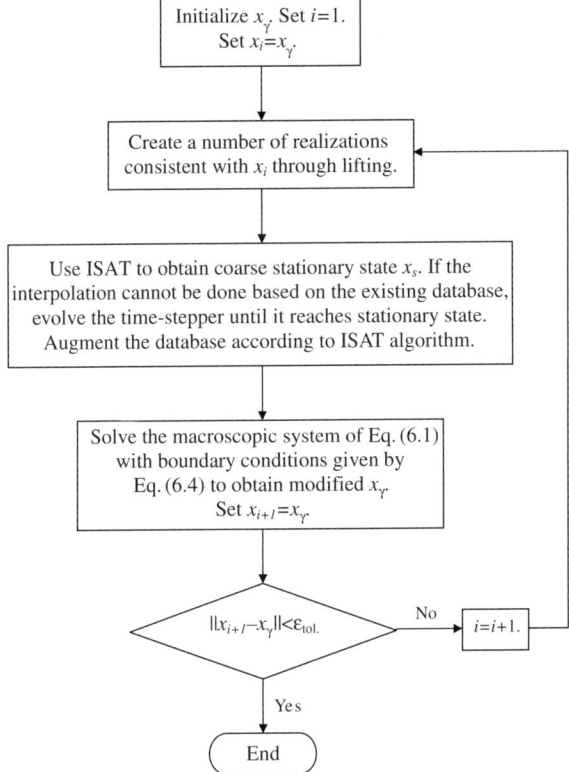

Fig. 6.2. Flowchart of the multiscale solution algorithm.

model in Eqs. (6.1)–(6.4) in the optimization problem in Eq. (6.7) to formulate an approximate optimization problem that can be solved using the standard search algorithms such as successive quadratic programming (SQP), BFGS, Luus–Jaakola, Hooke–Jeeves, derivative-free optimization, etc. [83, 22] to obtain the optimal solution. The proposed formulation leads to computational savings, while the accuracy of the solution remains close to the solution of the original optimization problem.

6.6 Application to Thin-Film Growth

6.6.1 Process Modeling and Simulation

The proposed optimization methodology is applied to a conceptual thin-film growth process inspired by GaN deposition, where the objective is to compute optimal time-invariant and time-varying process operating conditions that simultaneously minimize spatial thickness nonuniformity and surface roughness of the deposited film at the end of the process cycle.

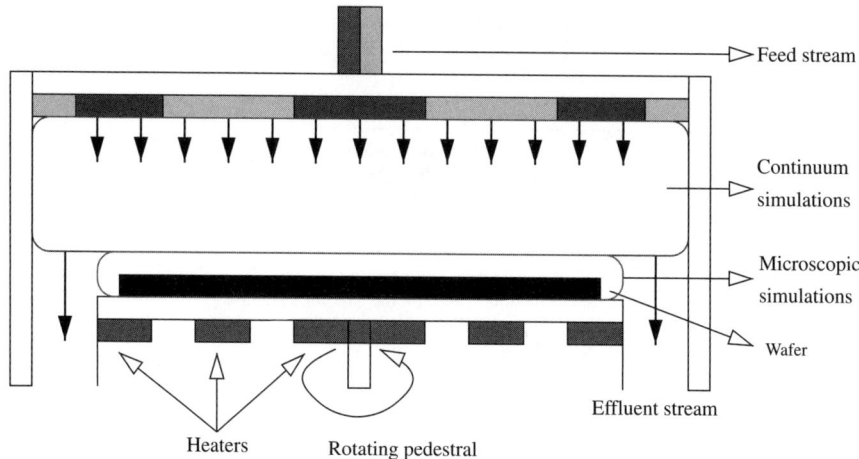

Fig. 6.3. Schematic of the reactor with a split-inlet showerhead configuration.

Figure 6.3 depicts the schematic of the reactor with a split-inlet showerhead configuration. The bulk of the reactor is modeled using two-dimensional axisymmetric PDEs in cylindrical coordinates derived from continuum conservation principles. The surface of the growing film is modeled using kMC simulations to compute the roughness of the growing film. Figure 6.3 shows the domains of description of the macroscopic and microscopic process models. It should be noted that the microscopic domain is infinitesimally thin. Substrate temperature profiles are manipulated using three circular heaters, with heat conduction being the prevailing heat transfer mechanism in between areas. Table 6.1 tabulates the reactor geometry and process conditions.

Table 6.1. Process conditions and reactor geometry.

Reactor radius	2 in.
Substrate radius- (R_s)	1.5 in.
Number of inlets	3
Inner inlet outer radius	0.5 in.
Middle inlet outer radius	1 in.
Outer inlet outer radius	1.5 in.
Substrate to inlet distance- (z_0)	3 in.
Reactor pressure	0.1 atm
Inlet & reactor wall temperature	300 K
Inlet temperature	300 K
Inlet velocity	80 cm/s
Substrate temperature- (T_s)	900–1300 K
Inlet mass fraction of species A (X_A)	0.4×10^{-2}
Inlet mass fraction of species B (X_B)	0.6

Table 6.2. Process reaction scheme.

Reaction	$k0$	E
(G1) $A \rightarrow A' + C$	1×10^{14}	39.9
(S1) $A' \rightarrow a(s) + D$	—[*]	0
(S2) $B \rightarrow b(s)$	—	—

[*] Rate calculated from kinetic theory of gases.

Gaseous species A and B represent the precursors of a and b (components of compound semiconductor ab), respectively, and are assumed to undergo the following gas-phase reactions in the bulk of the reactor and gas-surface reactions on the wafer surface, shown in Table 6.2.

Reaction G1 represents the thermal decomposition of precursor A into A', which adsorbs on the substrate (reaction S1). The rate parameter for adsorption of A' (reaction S1) is assumed to follow that of an ideal gas, i.e., $k_a = s_0 \sqrt{RT/2\pi M}$, where s_0 is the sticking coefficient. The adsorption rate of B (reaction S2) is assumed to be equal to S1 so that the stoichiometry of the film is preserved. In addition to adsorption, diffusion and desorption of adsorbed species are other significant processes that affect the structure of the surface. The desorption rate of the surface species into the gas phase and the rate of surface diffusion are calculated as

$$k_d^n = k_{d0} e^{-\frac{E_{d0}+n\Delta E}{k_B T}}, \quad k_m^n = \frac{k_B T}{h} e^{-\frac{E+n\Delta E}{k_B T}}, \qquad (6.40)$$

where h is Planck's constant, E and E_{d0} are the energy barriers for surface diffusion and desorption, respectively, ΔE is the interaction energy between two neighboring adsorbed species, and $n \in \{0, 1, 2, 3, 4\}$ is the number of nearest neighbors. The values of E, E_{d0}, ΔE, and k_{d0} are taken as 2.5 eV, 2.5 eV, 0.5 eV, and 1×10^{13}, respectively.

The macroscopic description of the process under consideration is given by the following conservation equations:

$$\nabla \cdot (\rho \mathbf{u}) = 0, \quad \nabla \cdot (\rho \mathbf{u}\, \mathbf{u}) - \nabla \cdot \mathbf{T} - \rho \mathbf{g} = 0,$$

$$\nabla \cdot (\rho \mathbf{u} T)] = -\nabla \cdot \mathbf{q} - \sum_k h_k W_k \dot{\omega},$$

$$\nabla \cdot (\rho \mathbf{u} Y_k) = -\nabla \cdot \mathbf{j}_k + W_k \dot{\omega}_k; \quad k \in \{1, 2, 3, 4\},$$

$$\mathbf{j}_k = -D_k \rho \nabla Y_k - D_{T,k} \frac{\nabla T}{T},$$

$$(6.41)$$

where ρ is the gas-phase density, $\mathbf{u}$ is the fluid velocity vector, $\mathbf{T}$ is the stress tensor, C_p is the specific heat capacity, T is the temperature, $\mathbf{q}$ is the heat flux due to conduction, and h_k, W_k, and Y_k are the partial specific enthalpy, molecular weight, and mass fractions of gas species. $\dot{\omega}_k$ and $\mathbf{j}_k$ are the net production rate due to homogeneous reactions and the mass flux, respectively, of species k. D_k and $D_{T,k}$ in the flux equation correspond to mass diffusion and thermal diffusion coefficients, respectively.

The flux boundary condition at the deposition surface is given by [155]

$$\mathbf{j} = R_{ad} = k_a C_{A'}|_s - <k_d> f(C_{a.s}, T_s, w_{A'A'}),\tag{6.42}$$

where R_{ad} is the net rate of adsorption, T_s is the surface temperature, and $<k_d>$, $C_{A'}|_s$, and $C_{a.s}$ are the effective desorption rate, concentration of A' over the sub-strate, and $average$ surface concentration of adsorbed $a(s)$, respectively. Function f describes the influence of lateral interactions on the desorption rate, which cannot be ascertained without knowledge of the surface structure. We employ kMC to account for the surface structure and estimate the right-hand side of Eq. (6.42), which links the two levels of descriptions.

kMC approximates the solution of the stochastic master equation (see Chapter 2) through Monte Carlo sampling;

$$\frac{\partial P(\sigma, t)}{\partial t} = \sum_{\sigma'} W(\sigma', \sigma) P(\sigma', t) - W(\sigma, \sigma') P(\sigma, t),$$

where σ and σ' are system configurations, $P(\sigma, t)$ is the probability that the system is in state σ at time t, and $W(\sigma, \sigma')$ is the probability per unit time of transition from σ to σ'. It is assumed that at any instant, only a single event (out of all possible events) occurs, according to its relative probability. After each event, time is incremented by δt, given as

$$\delta t = -\frac{\ln\ r}{\sum_i \Psi_i} = -\frac{\ln\ r}{k_a N_T + \sum_{n=0}^4 k_{m,n} N_n},\tag{6.43}$$

where r is a random number between 0 and 1 and Ψ_i is the propensity function of event i. The summation in the denominator is carried over all possible events, and transition probabilities are adjusted after each event. N_T is the total number of surface sites and N_n is the number of sites with n nearest neighbors. The surface roughness is computed from:

$$\mathcal{R} = \frac{1}{2N_T} \sum_{i,j} (|h_{i+1,j} - h_{i,j}| + |h_{i,j-1} - h_{i,j}|$$
$$+ |h_{i-1,1} - h_{i,j}| + |h_{i,j+1} - h_{i,j}|),\tag{6.44}$$

where $h_{i,j}$ is the number of atoms adsorbed at the (i, j)th surface site.

The temperature of the substrate can be manipulated using the three circular heaters placed below it. Two distinct forms of actuation were investigated. Initially, it was assumed that the effect of heaters on the substrate temperature is localized to rings of negligible thickness in space, and heat conduction is the prevailing mecha-nism of heat transfer in areas between the heaters. The following equation represents the resulting substrate temperature profile (denoted as thin-ring actuation):

$$T_s(\mathbf{u}, r) = \left(u_1 + \frac{u_2 - u_1}{0.5R_0}r\right) B_1(r, 0, R_0/2)$$
$$+ \left(u_2 + \frac{u_3 - u_2}{0.5R_0}\right)(r - 0.5R_0)B_1(r, R_0/2, R_0)\tag{6.45}$$
$$+ u_3 B_1(r, R_0, R_w),$$

where $B_1(\cdot)$ denotes the standard Boxcar function [defined in terms of Heaviside functions, $H(\cdot)$, as $B_a(x, b, c) = a[H(x - b) - H(x - c)]]$. Subsequently, the assumption of actuation being localized to rings of negligible thickness was relaxed to enforcing specific substrate temperatures over rings of finite thickness, giving rise to the following profile (denoted as ring actuation):

$$
\begin{aligned}
T_s(\mathbf{u}, r) = {} & u_1 B_1(r, 0, r') + u_2 B_1(r, R_w/2 - r'/2, R_w/2 + r'/2) \\
& + u_3 B_1(r, R_w - r', R_w) \\
& + \left(u_1 + \frac{u_2 - u_1}{R_w/2 - r'/2} r \right) B_1(r, r', R_w/2 - r'/2) \\
& + \left(u_2 + \frac{u_3 - u_2}{0.5R_w - 1.5r'}(r - R_w/2 + r'/2) \right. \\
& \qquad \left. \times B_1(r, R_w/2 + r'/2, R_w - r') \right)
\end{aligned}
\tag{6.46}
$$

For the generation of snapshots, the macroscopic domain, Ω_1, was discretized using finite differences into 6,201 nodes and the resulting system of nonlinear algebraic equations was solved using a Newton–Krylov-based solver. The finite-difference grid also partitioned the interface γ into 60 nodes. Simulation of the entire substrate surface using kMC was a computationally intractable task; hence, the gaptooth discretization scheme was initially investigated [55], linking the patches of kMC simulations. However, the intercommunication between the patches was found to be negligible due to the large distance between them. As a result, independent simulations were performed at the center of each discretization node assuming periodic boundary conditions. Depending on the structure of the kMC simulator, the flow of information across the interface of the continuum and the discrete domains can be unidirectional or bidirectional. For the current case, numerical simulations established that inclusion of desorption into the kMC model had a negligible effect on the macroscopic solution of the multiscale system. Hence, in the reduced-order process, model desorption was not included. Under this assumption, the flow of information was unidirectional and did not require multiple iterations.

6.6.2 Time-Constant Process Operation: Optimization Problem Formulation and Results

We initially consider the problem of time-constant process operation where precursor A flows through the innermost inlet and B through the two outer inlets (which is denoted as the ABB configuration of inlet), and the optimization objective is to determine surface temperature profiles that are optimal with respect to the cost functional. We formulate the optimization problem as

$$\min \; F = w_1 \int_0^{R_0} \left\{ R_{\text{dep}}(r) - \bar{R}_{\text{dep}} \right\}^2 dr + w_2 \int_0^{R_0} \mathcal{R}(r) dr$$

$$\text{s.t.}$$

$$R_{\text{dep}} = k_a C_{A'} \; \text{at} \; \gamma, \tag{6.47}$$

$$\bar{R}_{\text{dep}} = \frac{1}{R_0} \int_0^{R_0} R_{\text{dep}}(r) dr,$$

$$900 \leq T_s(u_k, r) \leq 1300,$$

where F is the objective functional, R_{dep} is the deposition rate of the thin film, $\mathcal{R}$ is the surface roughness of the deposited film, and T_s is the surface temperature. R_0 is the cutoff radius, which is taken to be a fraction of the substrate radius, thus discounting the unavoidable edge effects. The objective function penalizes nonuniformities in the deposition rate (macroscopic objective) and the surface roughness across the substrate, as well as high values of the spatially averaged roughness of the film (microscopic objectives). Additional constraints on the optimization problem arise from the reduced-order process model in Eq. (6.1), whose explicit form is omitted for brevity. The temperature of the substrate is the design variable in the above optimization problem and is manipulated using the three circular heaters placed below the substrate. Since two distinct forms of actuation were investigated, two independent optimization problems were considered, based on the actuators chosen, which were solved using a Hooke–Jeeves search algorithm [83].

The choice of the lattice size for kMC is an important trade-off (similar to the choice of grid size in the case of finite elements), since large lattice sizes increase the computational demand. On the other hand, kMC with a small domain size has significant stochastic uncertainty. Figure 6.4 compares the time evolution of surface roughness computed using 50×50, 75×75, and 100×100 lattices. It can be seen that the results from all the three lattice sizes are comparable. Hence, we chose a 75×75 lattice for the rest of the computations. The pertinent simulation parameters are listed in the figure's caption. During the development of the multiscale process model, it was assumed that a stationary state for coarse variables, $\bar{x}_s$, exists and is independent of the initial condition. In order to demonstrate the validity of this assumption with respect to surface roughness, Fig. 6.5 plots the time evolution of surface roughness for three distinct initial surface configurations. It can be seen that after a short transient, all three trajectories converge to a common stationary point. The caption to Fig. 6.5 contains the pertinent simulation parameters.

For the two cases, a separate ensemble of 729 snapshots was generated by varying u_1, u_2, and u_3 between 900–1300 K. Application of KLE generated 3, 25, and 40 empirical eigenfunctions for the temperature and mass fractions of A and A', respectively, for the former case (thin-ring actuation) and 3, 26, and 41 for the latter case (ring actuation). In both cases, the eigenfunctions accounted for more than 99.999% of the energy of the respective ensembles. The reduced-order model was hence comprised of 68 and 70 (as opposed to $6,201 \times 3$) nonlinear algebraic equations for the former and latter cases, respectively.

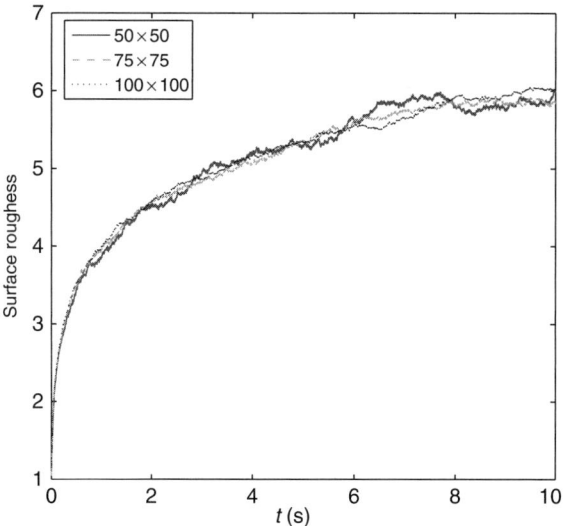

Fig. 6.4. Surface roughness as a function of time for various lattice sizes ($T_s = 1100\,\text{K}$; adsorption rate $= 10\text{atoms/site-s}$).

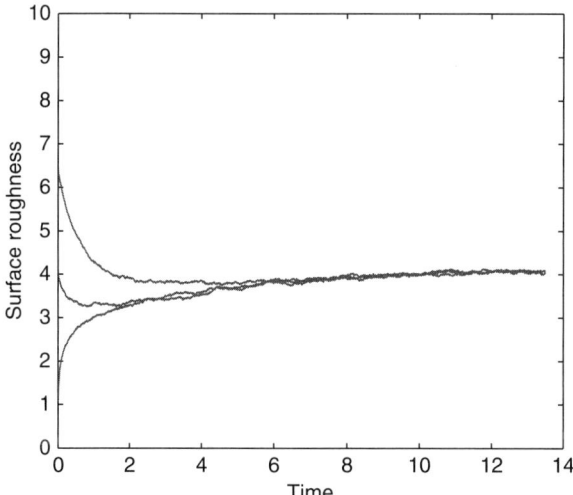

Fig. 6.5. Surface roughness as a function of time with different initial surface configurations ($T_s = 1150\,\text{K}$; adsorption rate $= 8\text{atoms/site-s}$).

Figure 6.6 presents surface deposition rate profiles of species a for $T_s = 900\,\text{K}$ and $T_s = 1300\,\text{K}$ and compares them against that of optimal surface temperature profiles, corresponding to thin-ring actuation, obtained by solving the problem in Eq. (6.47) by accounting first only for the macroscopic objective (i.e., $w_2 = 0$), and, second for the combined macroscopic-microscopic (multiscale) objective (i.e.,

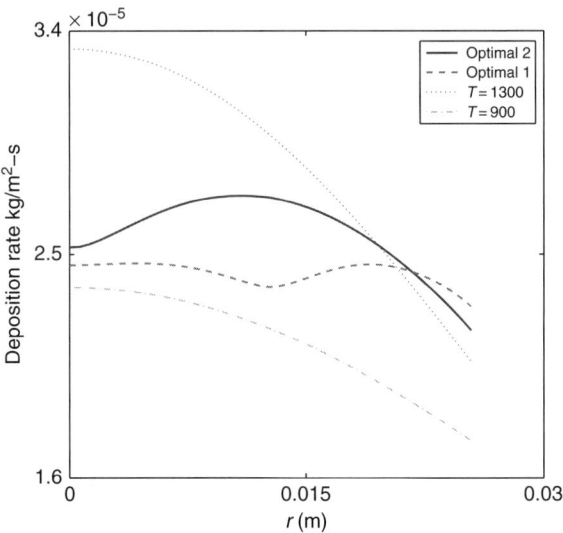

Fig. 6.6. Comparison of deposition rate profiles with macroscale-only (Optimal 1) and multiscale optimization objective (Optimal 2) (time-constant process operation).

$w_1, w_2 \neq 0$). Deposition rate nonuniformity is defined as

$$[\max_{r \in [0, R_0]} (R_{\text{dep}}(r)) - \min_{r \in [0, R_0]} (R_{\text{dep}}(r))]/ \max_{r \in [0, R_0]} R_{\text{dep}}(r)$$

and is computed to be 37.86%, 26.05%, 7.06% and 19.77% for $T_s = 1300\,\text{K}$, $T_s = 900\,\text{K}$ and the two optimal cases, respectively (see Table 6.3). The corresponding optimal temperature and surface roughness profiles are shown in Fig. 6.7 and 6.8, respectively. For constant substrate temperature operation, the adsorbate concentration, and, hence, the deposition rate, reduces across the substrate radius, except near the edge, where, due to high convective mass transfer, the deposition rate increases sharply. This effect is omitted in the computation of the deposition rate uniformity. However, temperature gradients near the substrate induced by radial

Table 6.3. Nonuniformity and roughness values for different operating policies

Operation	Objective	Nonuniformity	Roughness	Temperature profile
1300 K, constant	—	37.86%	1.95	
900 K, constant	—	26.05%	11.69	
Time-constant, thin-ring actuation	Macroscale	7.06 %	9.26	Fig. 6.7
Time-constant, thin-ring actuation	Multiscale	19.77%	6.59	Fig. 6.7
Time-constant, ring actuation	Macroscale	16.2%	11.0	Fig. 6.11
Time-constant, ring actuation	Multiscale	21.83%	7.8	Fig. 6.11
Time-varying, ring actuation	Macroscale	0.62%	3.5	Fig. 6.13
Time-varying, ring actuation	Multiscale	0.92%	1.7	Fig. 6.13

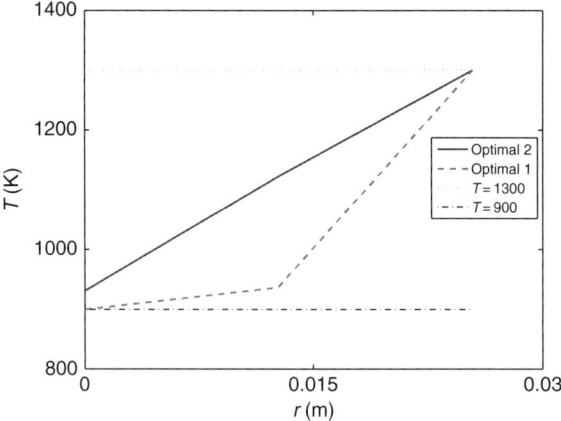

Fig. 6.7. Optimal substrate temperature profiles across the wafer surface (time-constant process operation).

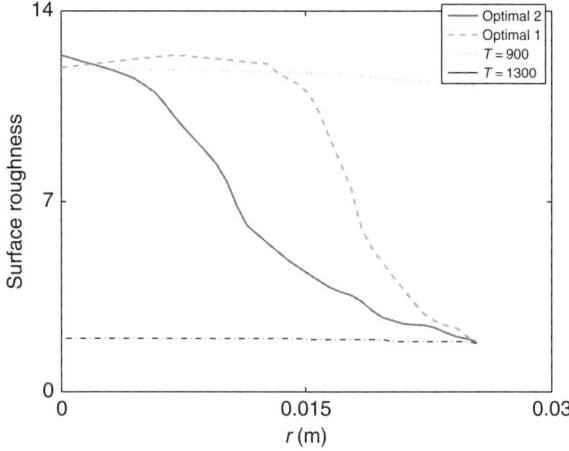

Fig. 6.8. Surface roughness profiles with macroscale-only (Optimal 1) and multiscale optimization objective (Optimal 2) (time-constant process operation).

variations in substrate temperature cause preferential diffusion of species radially, which leads to modified deposition rates. The solution of the optimization problem demonstrates that optimal variation of the substrate temperature (by controlling the magnitude of actuation) can lead to significant improvement in the uniformity of the grown film. Higher overall surface temperature is required if reduction of average surface roughness is a concurrent optimization objective (see Fig. 6.7) because at high temperatures, an increase in diffusion over the surface of the adsorbed species leads to relatively smoother films. However, the modified temperature gradients (especially near the substrate) may not favor axial thermal diffusion to the desired extent and, consequently, deposition rate uniformity may be compromised.

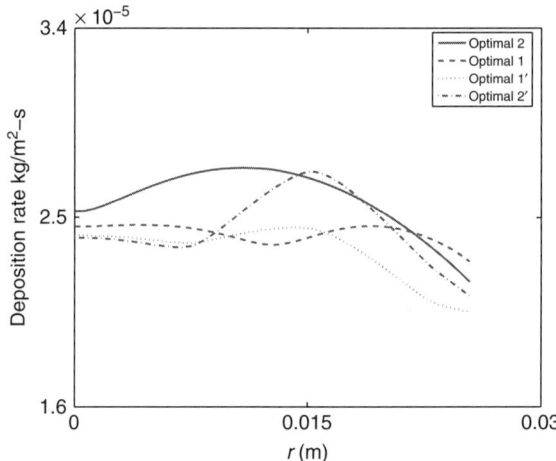

Fig. 6.9. Comparison of optimal deposition rate profiles with thin-ring and ring actuation for time-constant process operation. Optimal 1 and 1′ correspond to macroscale optimization with thin-ring and ring actuation, respectively. Optimal 2 and 2′ correspond to multiscale optimization.

The optimal temperature profile with respect to the above concurrent objective reduces the average surface roughness from 9.26 (for macroscale-only case) to 6.59, however, increasing the nonuniformity in deposition rate from 7.06% to 19.77%. It should be noted that process operation with substrate temperature near 1300 K results in a notable reduction in surface roughness, but, it is not optimal with respect to the deposition rate uniformity.

In order to account for the finite surface area of heaters, temperature actuation given by Eq. (6.46) was employed; the corresponding optimal surface temperature profiles for macroscopic-only and multiscale process objectives are shown in Fig. 6.9 and the corresponding roughness profiles are plotted in Fig. 6.10. The computed optimal deposition rate nonuniformity and average surface roughness of the film are 16.2% and 11 for the microscale objective and 21.83% and 7.8 for the multiscale objective, respectively. Optimal surface temperature profiles are shown in Fig. 6.11. Hence, process operation employing ring actuation resulted in compromised optimization objectives, which motivated the investigation of time-varying process operation.

6.6.3 Time-Varying Process Operation: Optimization Problem Formulation and Results

It is evident from the preceding subsection that optimization, based on time-constant process operation, resulted in considerable improvement of the quality of the thin film. However, the contrasting effect of the substrate temperature profile on deposition rate uniformity and surface roughness of the film posed limitations when

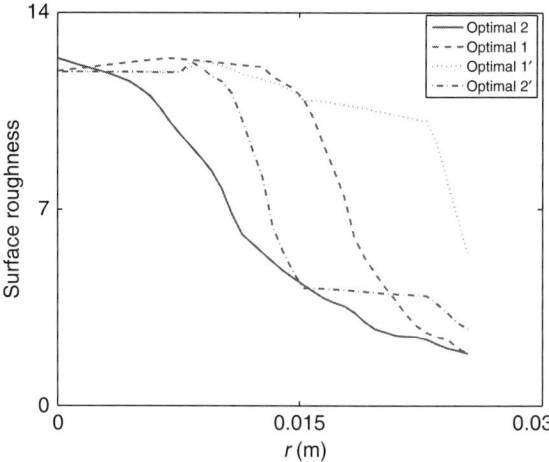

Fig. 6.10. Comparison of surface roughness profiles with thin-ring and ring actuation for optimal time-constant process operation. Optimal 1 and 1′ correspond to macroscale optimization with thin-ring and ring actuation, respectively. Optimal 2 and 2′ correspond to multiscale optimization.

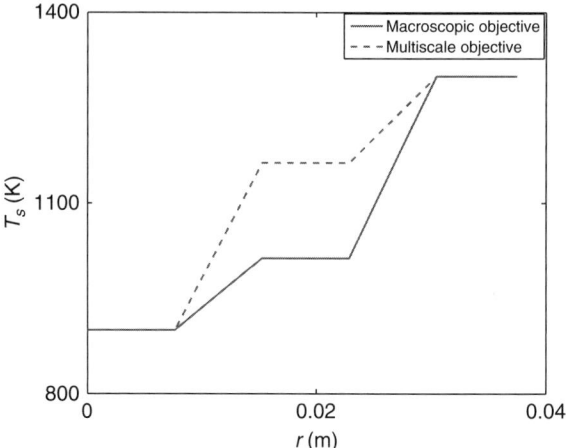

Fig. 6.11. Optimal temperature profiles for ring temperature actuation (time-constant process operation).

simultaneous macroscale and microscale process objectives were desired. This issue is addressed in this subsection by employing time-varying process operation. As mentioned earlier, for steady-state process operation, precursor A flows through the innermost inlet and B through the two outer inlets (ABB configuration of inlet). During the process operation, the two gas streams in the two innermost inlets can be interchanged to result in a distinct (BAB) inlet configuration. The transient evolution of the process following the switch is neglected due to the long time associated between switches in comparison to the time needed for relaxation of the system dynamics due to switch (quasi-steady-state approximation). It is proposed that by optimally switching from an ABB to a BAB configuration and controlling the substrate temperature profile before and after the switching, both optimization objectives can be realized. Mathematically, the optimization problem can be formulated as

$$\min \quad F = \int_0^{R_0} \left\{ w_1 [\mathcal{T}(r) - \mathcal{T}_{\text{obj}}]^2 + w_2 \mathcal{R}(r) \right\} dr$$

$$\text{s.t.}$$

$$\mathcal{T} = \sum_{i=1}^{n} \delta \bar{t}_i R_{\text{dep}}, \qquad \delta \bar{t}_i = \bar{t}_{i+1} - \bar{t}_i,$$

$$R_{\text{dep}} = k_a C_{A'} \quad \text{at} \quad \gamma, \tag{6.48}$$

$$u_k = \sum_{i=1}^{n} u_{k,i} [H(\bar{t}_i) - H(\bar{t}_{i+1})], \quad k \in \{1, 2, 3\},$$

$$900 \leq T_s(u_k, r) \leq 1300,$$

where F is the objective functional, $\mathcal{T}$ is the thickness of film at the end of the deposition, $\mathcal{T}_{\text{obj}}$ is the target thickness of the film, R_{dep} is the deposition rate of species a, $\mathcal{R}$ is the surface roughness of the deposited film, $\delta \bar{t}_i$ is the time interval for ith switching, T_s is the surface temperature, and u_k is the magnitude of actuation. The objective function penalizes any deviation of final film thickness from the target thickness (macroscopic objective) and high values of the spatially averaged roughness of the film (microscopic objectives). The design variables of the optimization problem are the magnitudes of actuation u_k and the time intervals $\delta \bar{t}_i$.

The optimization problem in Eq. (6.48) for the time-varying process operation was solved in two steps. Initially, spatial uniformity of the deposited film was the only optimization objective (i.e., $w_1 \neq 0, w_2 = 0$). Subsequently, the microscopic objective was included in the optimization. The target film thickness, $\mathcal{T}_{\text{obj}}$ was 5×10^{-6} m. Surface temperature profiles represented by the actuation of Eq. (6.46) were employed, with conduction being the prevailing mechanism of heat transfer between the actuators.

An ensemble of solution data ("snapshots") was generated by varying u_1, u_2, and u_3 for both the ABB and BAB inlet configurations and solving the resulting system according to the algorithm presented earlier. Specifically, an ensemble of 729×2 snapshots was generated. Three, 62, and 52 eigenfunctions were identified using KLE, respectively, for temperature and mass fraction profiles of A and A' across the reactor, which captured more than 99.999% of the ensemble's energy. The reduced-order model was hence comprised of 117 (as opposed to $6,201 \times 3$) nonlinear algebraic equations. Coarse data of kMC simulations were tabulated in accordance with in situ adaptive tabulation, as described earlier, which facilitated efficient linking between the macroscopic and microscopic process models.

Figure 6.12(a) shows the final film thickness across the wafer surface obtained for the optimal process operation with macroscopic objective and multiscale objective

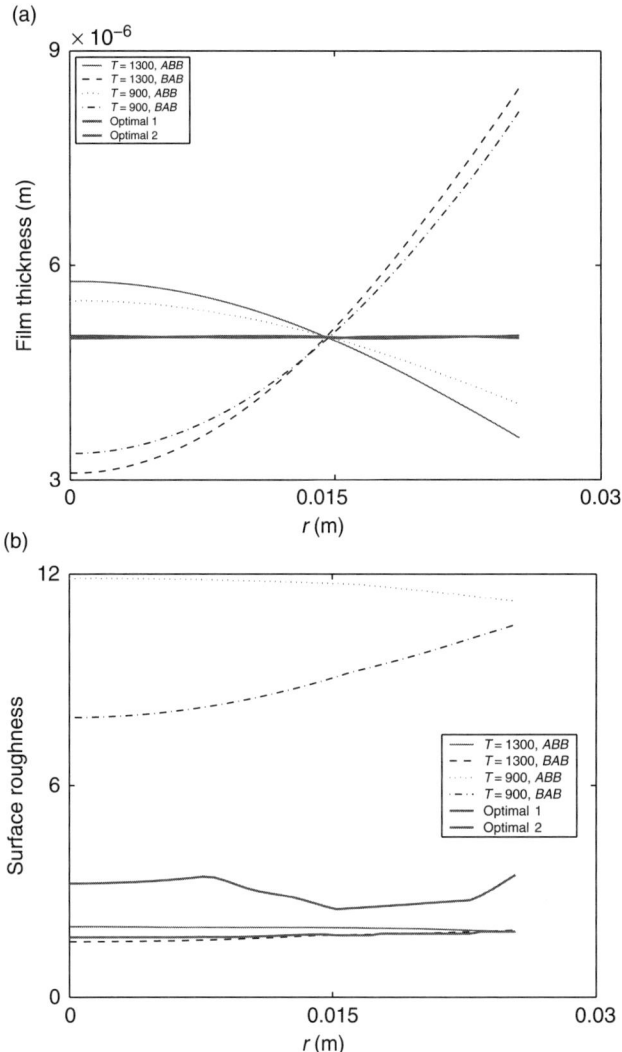

Fig. 6.12. Comparison of (a) deposition rate and (b) roughness profiles across the wafer surface, with macroscale (Optimal 1) and multiscale objective (Optimal 2) (time-varying process operation).

(denoted as Optimal 1 and Optimal 2, respectively). For comparison purposes, the final film thickness profiles for time-invariant nominal process operation are also shown. Thickness nonuniformity, defined as

$$[\max_{r\in[0,R_0]}(\mathcal{T}(r)) - \min_{r\in[0,R_0]}(\mathcal{T}(r))]/\max_{r\in[0,R_0]}\mathcal{T}(r)$$

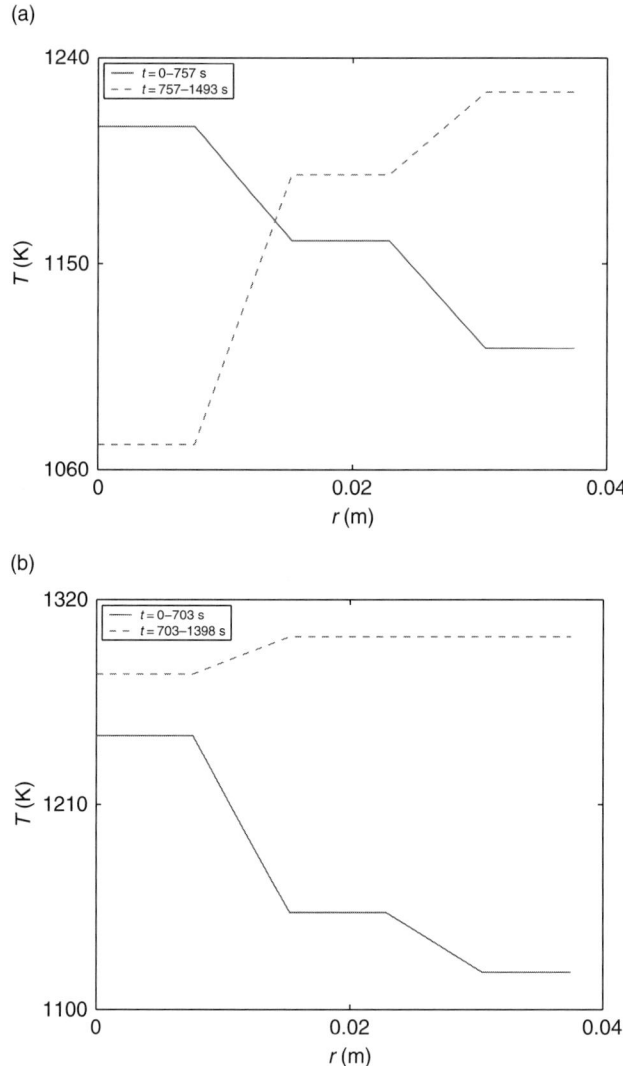

Fig. 6.13. Initial (solid lines) and final (dashed lines) optimal surface temperature profiles, for (a) macroscale-only objective and (b) multiscale objective (time-varying process operation).

for the ABB inlet configuration, was found to be 37.86% and 26.05%, respectively, for the substrate temperatures 1300 K and 900 K. The corresponding numbers were 63.45% and 58.63%, respectively, for the BAB inlet configuration. However, under the optimal process operation, radial nonuniformity in the film was reduced to 0.62% and 0.92% for the former and latter cases, respectively. The corresponding inlet-switching policies and substrate temperature profiles before and after switching are shown in Fig. 6.13(a) and 6.13(b).

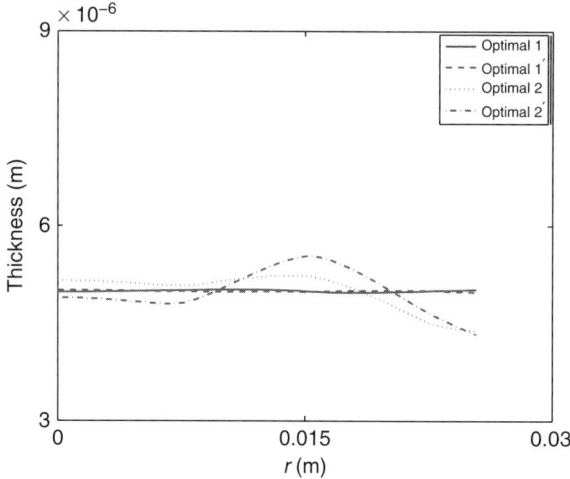

Fig. 6.14. Comparison of film-thickness profiles with time-constant and time-varying process operation. Optimal 1 and 1′ correspond to macroscale optimization with thin-ring and ring actuation, respectively. Optimal 2 and 2′ correspond to multiscale optimization.

Inclusion of the microscopic objective resulted in the overall increase of substrate temperature and the spatially averaged surface roughness of the film decreased from 3.5 (for Optimal 1) to 1.7 (for Optimal 2). Surface roughness profiles are shown in Fig. 6.12(b). It should be noted that the BAB configuration with $T_s = 1300\,\mathrm{K}$ would result in the film with the lowest surface roughness; however, such an operation is not optimal with respect to spatial thickness uniformity of the film.

Figure 6.14 compares the optimal final film thickness across the wafer for time-constant and time-dependent process operations. The improvement in process optimization objectives is clearly evident, implying that the time-varying operation is preferred over the time-constant operation, while the computational increase in the optimization is relatively small. The film-thickness uniformity improved from 21.83% to 0.92%, and the average surface roughness across the wafer improved from 7.8 to 1.7 by employing time-varying process operation in place of time-constant process operation.

Remark 6.8. The optimization problem for the time-dependent process operation was also solved for thin-ring actuation represented by Eq. (6.45). A considerable improvement in process objectives was observed in this case too. The film-thickness uniformity notably improved from 19.77% to 0.71%, and the average surface roughness across the wafer reduced from 6.59 to 1.8 by employing time-varying process operation in place of time-constant process operation. Since these results were only marginally different from those obtained from distributed actuation of the substrate, their graphical presentation was omitted for brevity.

Remark 6.9. It is worth mentioning that random and pattern-search methods (such as Hooke–Jeeves) are not very efficient when the number of optimization parameters is large, which is typically the case if one uses an infeasible path approach. In the current context, if a full-order infeasible path approach is employed, the multiscale process model appears as equality constraints in the optimization problem. The number of equality constraints, equal to the degrees of freedom, arising due to the macroscopic process model, are determined by the discretization of the process variables in the spatial domain. The extent of discretization can be of the order of 10^3–10^6 depending on the desired level of accuracy. Microscopic time-steppers, as outlined in Section 6.6.1, cannot be readily incorporated into the infeasible path approach. Additional degrees of freedom can be a part of the operating and process parameters, which in the present study were three (magnitude of temperature actuation) for the time-constant process operation and eight (magnitudes of temperature actuation and switching time) for the time-varying process operation. Since we employed the feasible path approach, the number of degrees of freedom was small (and one does not anticipate efficiency issues as far as the optimization method is concerned). The bottleneck for optimization is the computational complexity of the process model, which we propose to circumvent using the proposed order-reduction strategy.

Finally, the computational requirements of the proposed scheme are tabulated in Table 6.4 for the various optimization problems discussed in the previous sections. The CPU requirements listed are for a Pentium IV 3.02 GHz processor. It should be noted that the time required for the construction of ensemble of snapshots was not included in the CPU time reported in Table 6.4.

Table 6.4. Optimization efficiency.

Operation	Objective	Solution Time (min)	Function Calls
Time-constant, thin-ring actuation	Macroscale	15	131
Time-constant, thin-ring actuation	Multiscale	18	159
Time-constant, ring actuation	Macroscale	11	99
Time-constant, ring actuation	Multiscale	19	158
Time-varying, ring actuation	Macroscale	286	1012
Time-varying, ring actuation	Multiscale	246	1011

6.7 Conclusions

In this chapter, nonlinear order-reduction techniques for elliptic partial differential equations were employed together with the adaptive tabulation method for microscopic/mesoscopic simulators to derive reduced-order continuum-discrete hybrid models for the computationally efficient solution of optimization problems when the cost function and/or equality constraints necessitate the consideration of phenomena that occur over widely disparate length scales. The proposed methodology was applied to a conceptual thin-film growth process to simultaneously maximize the deposition rate uniformity and minimize the average surface roughness of the film. Optimal surface temperature profiles were calculated for time-constant process operation and optimal surface temperature and precursor inlet concentration trajectories were computed for time-dependent process operation. The use of reduced-order process model resulted in considerable savings in wall-clock time for optimization, which was virtually intractable using the complete PDE/kMC process model.

Dynamic Optimization of Multiscale Process Systems

7.1 Introduction

In this chapter, we extend the methodology outlined in the previous chapter towards the efficient solution of dynamic optimization problems for multiscale processes coupling continuum and discrete descriptions. The approach relies on the reduction of the continuum system using Galerkin–KLE and of the microscopic simulator using ISAT. The reduced systems are linked together to formulate a computationally efficient multiscale model (with bidirectional flow of information). Consequently, standard search algorithms can be employed for the solution of the optimization problem. The approach is demonstrated on two numerical examples describing catalytic oxidation of CO to CO_2. In the first case, lateral interactions between adsorbed species are neglected and the mobility of the adsorbed species is assumed to be infinitely fast, so that the system can be approximated as a well-mixed system. In the latter case, nearest-neighbor interactions between adsorbed CO molecules are introduced along with the finite mobility of the adsorbed species. The microscopic system is linked to a macroscopic system describing the diffusion of CO and O_2 on the catalyst surface. In both cases, optimal inlet concentration profiles are computed to guide the microscopic system from one stable stationary state to another stable stationary state. The results of this chapter were first presented in [152].

7.2 Problem Formulation

We focus on spatially distributed multiscale processes with the following state-space description:

$$\frac{\partial x}{\partial t} = \mathcal{A}(x) + f(t, x, u), \quad x(z, 0) = x_0(z), \quad \text{on } \Omega_1, \tag{7.1}$$

$$x_m(t_i) = \Pi(x_m(t_{i-1}), \delta t, x(\gamma, t)), \quad \text{on } \Omega_2, \tag{7.2}$$

$$\delta t = t_i - t_{i-1},$$

P.D. Christofides et al., *Control and Optimization of Multiscale Process Systems*,
Control Engineering, DOI 10.1007/978-0-8176-4793-3_7,
© Birkhäuser Boston, a part of Springer Science+Business Media, LLC 2009

$$g\left(x, \frac{dx}{d\eta}, u, \bar{x}\right) = 0, \quad \text{on } \Gamma, \tag{7.3}$$

$$\bar{x} = L(x_m). \tag{7.4}$$

Equations (7.1) and (7.2) represent the macroscopic and microscopic descriptions of the process over the respective domains Ω_1 and Ω_2. It is assumed that Ω_1 and Ω_2 do not overlap, that they share a common interface $\gamma(\subseteq \Gamma) = \Omega_1 \cap \Omega_2$, and that $\Omega = \Omega_1 \cup \Omega_2$ spans the whole process domain. $x(z,t) \in \mathbb{R}^n$ denotes the vector of macroscopic state variables, $t \in [0, t_f]$ is the time (t_f is the terminal time), $z = [z_1, z_2, z_3] \in \Omega_1 \subset \mathbb{R}^3$ is the vector of spatial coordinates, Ω_1 is the domain of definition of the macroscopic process, and Γ is its boundary. $\mathcal{A}(x)$ is a dissipative, possibly nonlinear, spatial differential operator that includes up to second-order spatial derivatives, $f(t, x, u)$ is a nonlinear, possibly time-varying, vector function that is assumed to be sufficiently smooth with respect to its arguments, $u \in \mathbb{R}^p$ is the vector of design variables that are assumed to be piecewise-continuous functions of time, and $x_0(z)$ is a smooth vector function of z.

Function Π, which describes the microscopic system, is a time-stepper, which interacts via an input–output structure and may be unavailable in closed form. It uses $x_m(t_{i-1})$ and the macroscopic state at the interface γ as input, evolves over the time interval δt, and produces state $x_m(t_i)$. The vector function $h(x, dx/d\eta, u, \bar{x})$ represents the boundary conditions at Γ, and $\left.\frac{dx}{d\eta}\right|_\Gamma$ denotes the derivative in the direction perpendicular to the boundary. The boundary conditions depend upon $\bar{x}$, the "coarse" realization of the microscopic state, thus linking the macroscopic system with the microscopic system. Coarse variables are defined through the restriction operator denoted as $L(\cdot)$. Typically, they are lower-order statistical moments of microscopic states chosen such that the coarse dynamics are observable.

Assumption 7.1. *There exists an invertible transformation* $y = k(x, dx/dz, u, \bar{x})$ *such that* $g(k^{-1}(y), k^{-1}(y)/d\eta, u, \bar{x}) = h(y, dy/d\eta) = 0$.

With this assumption, the system in (7.1) and (7.3) can be written as

$$\frac{\partial y}{\partial t} = k_x \mathcal{A}(k^{-1}y) + k_x f(t, k^{-1}y, u) + k_u \dot{u} + k_{\bar{x}} \dot{\bar{x}}, \tag{7.5}$$

$$y(z, 0) = k\left(x_0(z), \frac{dx_0}{dz}, u(0), \bar{x}(0)\right), \quad h\left(y, \frac{dy}{d\eta}\right) = 0. \tag{7.6}$$

A general optimization problem for the multiscale system in (7.1)–(7.4) can be formulated as

$$\min_{u(t)} \mathcal{G} = \int_0^{t_f} \int_\Omega G(x, \bar{x}, u) dz \, dt + W(x(t_f), \bar{x}(t_f))$$

s.t.

$$\frac{\partial x}{\partial t} = \mathcal{A}(x) + f(t, x, u), \quad x(z, 0) = x_0(z), \quad \text{on } \Omega_1,$$

$$x_m(t_i) = \Pi(x_m(t_{i-1}), \delta t, x(\gamma, t)), \quad \delta t = t_i - t_{i-1}, \quad \text{on } \Omega_2, \qquad (7.7)$$

$$g\left(x, \frac{dx}{d\eta}, u, \bar{x}\right) = 0, \quad \text{on } \Gamma, \quad \bar{x} = L(x_m),$$

$$p(x, \bar{x}, u) \le 0,$$

where $\mathcal{G}(\cdot)$ is a measure of the process performance at both macroscopic and microscopic scales, the function $W(\cdot)$ is the terminal cost, t_f denotes the terminal time, and $p(x, \bar{x}, d)$ is the vector of inequality constraints, which may include bounds on state and control variables. Due to the stochastic nature of the function Π and its unavailability in closed form, standard gradient-based algorithms are unsuitable for the solution of the above optimization problem. Also, it is inefficient to directly apply derivative-free search algorithms that compute the objective function as a black box due to the computational intensity of the multiscale model. To address this problem, we present the following reduced-order formulation of the above multiscale model; it employs a proper orthogonal decomposition for reduction of the PDEs and ISAT for efficient simulation of the microscopic time-stepper.

7.3 Multiscale Solution Algorithm

The reduced-order multiscale model is constructed by appropriately linking the reduced-order macroscopic description obtained from KLE with the microscopic description implemented within the framework of the ISAT algorithm. To ensure consistency across both length scales, we employ the "lift-evolve-restrict" scheme [53] outlined in the following steps:

1. Initialize macroscopic state, $x(0)$, and coarse microscopic state, $\bar{x}(0)$. Let x_Γ denote values of macroscopic variables at the interface Γ, and let $\mathcal{D}$ denote the ISAT database. Define $\phi = [\bar{x}, x_\Gamma]^T$. Initialize $t = 0$.
2. Integrate the macroscopic system from $x(t)$ by a time step τ to obtain $x(t + \tau)$ and $x_\Gamma(t + \tau)$.
3. If $\phi(t)$ is within the EOA of one of the nodes in $\mathcal{D}$, interpolate $\phi(t + \tau)$ from the database. Otherwise,
   (a) Transform $\bar{x}(t)$, through *lifting*, to a number of consistent microscopic realizations, $x_m(t) = \mu \bar{x}(t)$, where μ denotes the lifting operator,
   (b) Evolve the realizations using the microscopic simulations, to obtain $x_m(t + \tau)$. The evolution will be a function of $x_\Gamma(t)$.
   (c) Obtain the values of $\bar{x}(t + \tau)$ from the restriction of $x_m(t + \tau)$.
4. Set $t = t + \tau$; go to step 2.

Although the multiscale simulation approach described above uses an explicit integration scheme, implicit variants can be implemented in a straightforward manner. The choice of time step, τ, is crucial as it should be sufficiently large to allow the incorrectly initialized statistics of the microscopic system during lifting to relax to their true values.

Using the above reduced-order formulation of the multiscale system, the dynamic optimization problem (7.7) can be efficiently solved using standard optimization algorithms. Due to the stochastic nature of the discrete system, it may be more advantageous to employ a pattern-search algorithm [95] in the current chapter we use such an algorithm. The objective function is computed as a black box during optimization. The overall scheme is presented in Fig. 7.1.

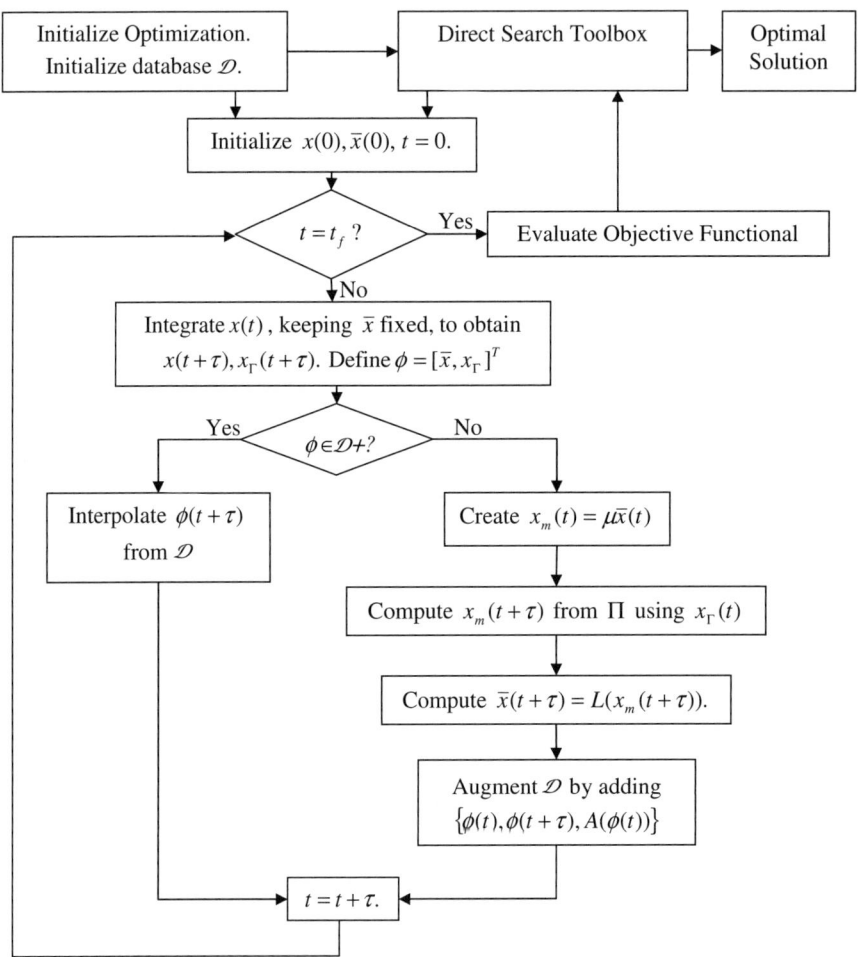

Fig. 7.1. Flowchart of the multiscale solution algorithm for the dynamic optimization of multiscale process systems.

7.4 Numerical Simulations

7.4.1 Catalytic CO Oxidation with Infinite Surface Mobility

We consider a kinetic model describing CO oxidation by O_2 on a catalytic surface [10]. The model involves transport of CO and O_2 to the catalyst surface, Langmuir adsorption for CO on the catalyst surface, dissociative adsorption of O_2 on the catalyst surface and second-order surface reaction to produce CO_2, which desorbs instantaneously. The overall surface reaction can be summarized as $A + 1/2B_2 \rightarrow C$, where A, B, and C represent CO, O and CO_2, respectively. The bulk transport of A and B is described by a set of parabolic PDEs, and the adsorption, desorption, and surface reaction phenomena are described by kMC simulations on a square lattice with $M \times M$ sites. The two levels of descriptions are linked through appropriate boundary conditions. The multiscale model is mathematically written as follows:

$$\frac{\partial x}{\partial t} = \frac{\partial^2 x}{\partial z^2}, \; z \in [0,1], \; x(z,0) = x_0,$$

$$x_m(t + \delta t) = \Pi(x_m(t), x(1,t)), \; x_m(0) = x_{m0},$$

$$x(0,t) = x^0(t), \; \frac{\partial x}{\partial z} + \lambda(x,\theta) = 0 \text{ at } z = 1, \qquad (7.8)$$

$$\theta_A = \int_{\Omega_2} ||x_m||_A d\Omega_2, \; \theta_B = \int_{\Omega_2} ||x_m||_B d\Omega_2,$$

where $x = [x_A, x_B]^T$ is the vector of gas-phase concentrations of A and B and represents the macroscopic variables, the microscopic state $x_m \in \mathbb{I}^{M^2}$ denotes the occupancy of the lattice sites, $\theta = [\theta_A, \theta_B]^T$ is the vector denoting surface coverage of adsorbed A and B and represents the coarse variables, and $\lambda(\cdot) = [x_A(t)(1 - \theta_A - \theta_B) - \gamma\theta_A, 2x_B(1 - \theta_A - \theta_B)^2]$ computes the net rates of adsorption of A and B. The function $|| \cdot ||_A$ is defined such that $||x_m(i)||_A = 1$ if the ith surface site is occupied by A and $||x_m(i)||_A = 0$ otherwise. It can be shown that the microscopic subsystem exhibits multiple steady states for a range of values of x_B^0.

We are interested in the solution to the following constrained dynamic optimization problem:

$$\min_{x_B^0(t)} F = Q(x,\theta,t) + W(x_{t_f}, \theta_{t_f}),$$

$$Q = \int_0^{NT} (x_B^0(t) - x_{B,ss}^0)^2 (1 - 0.3e^{-t}) T \sum_{i=1}^N \delta(t - iT)dt, \qquad (7.9)$$

$$W = 50[1 - e^{-R(|\theta_A(t_f) - \theta_{A,ss,f}| - \epsilon)} e^{-R(|\theta_B(t_f) - \theta_{B,ss,f}| - \epsilon)}].$$

The optimization objective is to compute an optimal boundary concentration profile, $x_B^0(t)$, such that the microscopic state of the system switches from an initial stable stationary state $\theta_{ss,i}$, to another stable stationary state, $\theta_{ss,f}$, within finite time $t_f = NT$ (see Table 7.1 for parameter values). The function Q evaluates the performance of the process over time period $[0,t_f]$ and the function W penalizes any deviation

Table 7.1. CO oxidation process parameters.

Parameter	Value	Steady states	
k_r	1.0	$\theta_{A,ss,i}$	0.13944
α	1.6	$\theta_{A,ss,u}$	0.67526
γ	0.04	$\theta_{A,ss,f}$	0.97101
$x_{B,ss}^0$	3.8	$\theta_{B,ss,i}$	0.63553
t_f	5s	$\theta_{B,ss,u}$	0.11452
		$\theta_{B,ss,f}$	0.00137

of the final state from the desired steady state. Due to the stochastic nature of the microscopic system, the final state is restricted to be in the neighborhood of the desired steady state and ϵ is a parameter for which $|\theta(t_f) - \theta_{ss}| \leq \epsilon \Rightarrow \mathcal{W} = 0$. θ is the standard ramp function.

The constraints for the optimization problem arise from the multiscale dynamic process model. To solve the optimization problem, the manipulated variable was parameterized into N time intervals of length T and the objective functional, F, was computed independently from the dynamic simulation of the system. A pattern-search algorithm [95] was employed for the solution of the above optimization problem. Initially, the problem was solved for $N = 10$ and $T = 0.5\,\text{s}$. For order reduction of the parabolic PDEs, an ensemble consisting of 500 snapshots of PDE solution data was initially constructed. Application of KLE resulted in four eigenfunctions each for species A and B that captured more than 99.99% of the energy contained within the ensemble. Microscopic simulations were performed using ISAT with $\phi = [\theta_A, \theta_B, x(1, t)]^T$. Note that with the choice of coarse variables $\bar{x} = [\theta_A, \theta_B]^T$, the microscopic dynamics are observable. Initially, the database was empty; it was concurrently built during the solution of the optimization problem. The reduced multiscale process model was created by linking the two descriptions using the algorithm presented in Section 7.3. The ISAT-reporting horizon was $\tau = 0.05\,\text{s}$, so that the database was queried 100 times per objective functional evaluation. This implies that in total 100 evaluations were need to compute the microscopic process trajectory per function evaluation. The optimal control trajectory obtained from the solution of the optimization problem is shown in Fig. 7.2. For comparison, the figure also presents the "deterministic" optimal trajectory, which is obtained by using the mean-field description of the microscopic system in the multiscale algorithm. The difference between the two profiles is attributed to the noise arising from the kMC simulations.

Figure 7.3 plots the number of database interpolations and number of time-stepper evaluations performed as a function of objective functional evaluation during optimization. It can be observed that database interpolations are significant during the majority of the objective functional evaluations, thereby speeding up the solution of the optimization problem. An average CPU time of 6 seconds was required to compute the objective functional using the reduced multiscale model, which is significantly smaller than the 21 seconds required using the full multiscale model. Note that the computational savings are a manifestation of model reduction at both

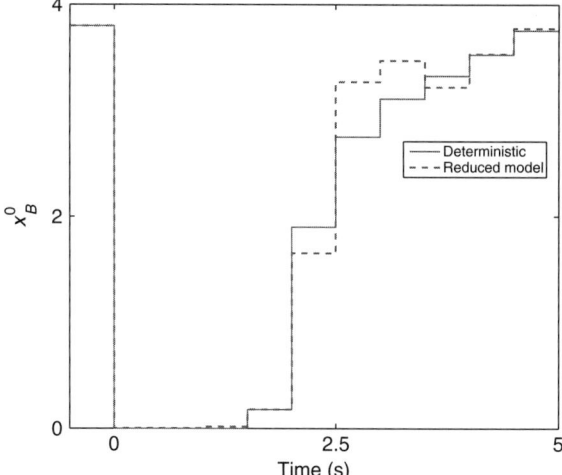

Fig. 7.2. Optimal control profile for $N = 10$.

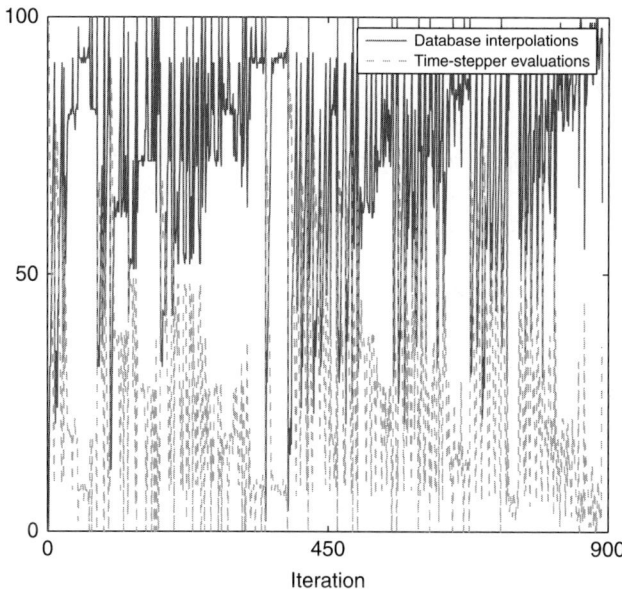

Fig. 7.3. Number of time-stepper evaluations and database interpolations per F computations for $N = 10$. Total equal to 100 per function evaluation.

macroscopic and microscopic scales. Greater computational savings can be expected as the database increases in size. The average number of time-stepper evaluations per F evaluation was 27, which is significantly lower than the 100 time-stepper evaluations required without interpolation.

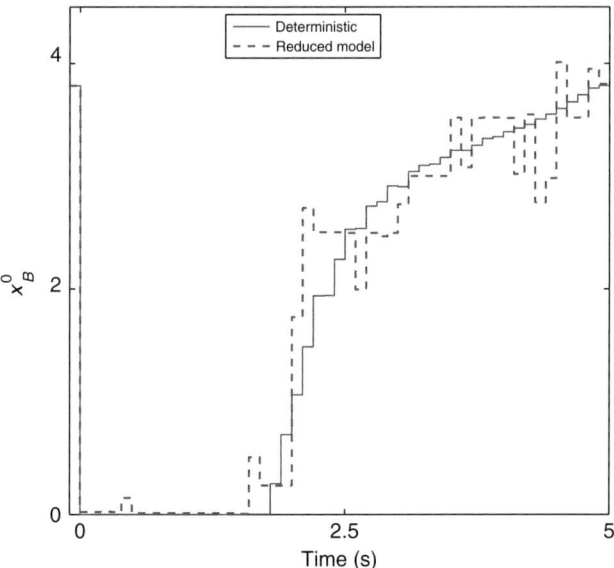

Fig. 7.4. Optimal control profile for $N = 50$.

Subsequently, the optimization problem was solved with $N = 50$ and $T = 0.1\,s$ to obtain an improved resolution in $x_B^0(t)$. The database created previously was employed as the initial database. The resulting optimal profile for $x_B^0(t)$ is shown in Fig. 7.4. The noisy behavior is a manifestation of the stochastic nature of the dynamic system resulting in performance deterioration of pattern-search algorithms. The number of database interpolations and number of time-stepper evaluations performed as a function of F evaluation during the optimization are plotted in Fig. 7.5. It can be seen that initially the number of time-stepper evaluations required for the computation of F is high, which, however, continuously decreases as the optimization proceeds. The advantage of using the preexisting database is clearly evident in this case, as the average number of time-stepper evaluations per F calculations reduced to 9 compared to 27 in the previous case. An average of 3.5 seconds was required to compute the objective functional using the reduced multiscale model, which is considerably shorter than the 21 seconds required using the full multiscale model.

7.4.2 Catalytic CO Oxidation with Limited Surface Mobility and Lateral Interactions

In the following section, we relax the earlier assumption that surface species possess infinite mobility and hence can be considered as a well-mixed system. Also, we incorporate first nearest-neighbor interactions between adsorbed CO molecules.

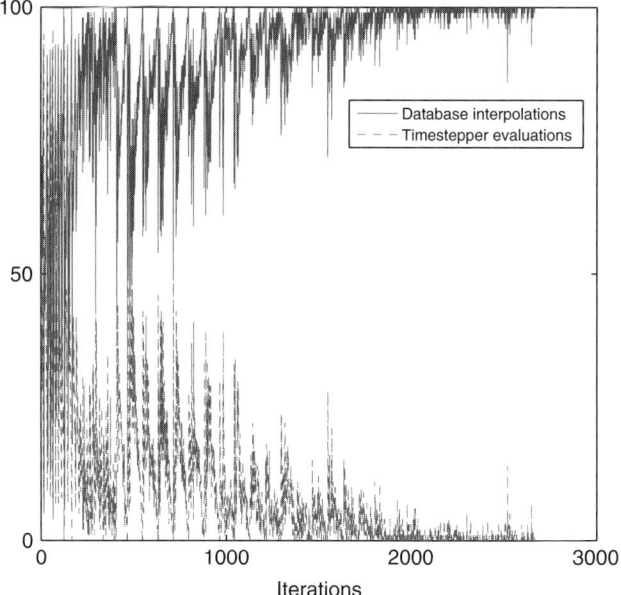

Fig. 7.5. Number of time-stepper evaluations and database interpolations per F computations for $N = 50$. Total equal to 100 per function evaluation.

Consequently, the rate coefficient for desorption of CO is now computed as a function of local environment, i.e., $\gamma \exp(-n\mathrm{E}/RT)$, where E denotes the lateral interaction strength between adsorbed CO molecules and n is the number of CO first nearest-neighbor CO atoms. In addition, the rates of surface migration are computed as $D_0 \exp(-n\mathrm{E}/RT)$, where D_0 is the migration rate at zero coverage of CO.

Unlike the previous case, exact closed-form equations for the evolution of surface coverage of O and CO do not exist for this case, although several levels of approximations have been obtained previously [141, 106]. These include modified mean-field equations to account for lateral interactions and quasi-chemical approximations based on pair probabilities. From the simulation point of view, it implies that microscopic simulations, initialized based on surface coverages during lifting, will only be a crude approximation. A better approximation is obtained if pair probabilities are also accounted for during lifting. This is based on the fact that the remaining higher-order moments of species distribution have faster dynamics and quickly become algebraic functions of the lower-order moments [53].

The optimization problem of Eq. (7.9) is resolved with the modified microscopic description incorporating lateral interaction for W = 10 and T = 0.5s. The relevant parameters are listed in Table 7.2. Note that, due to the lack of closed-form dynamical expressions, the optimization results cannot be compared with the corresponding "deterministic" results. Figure 7.6 shows the optimal input concentration trajectory obtained from the solution of the optimization problem employing a reduced multiscale process model. The figure also compares the above result with the

Table 7.2. CO oxidation with lateral interaction process parameters.

Parameter	Value	Steady states	
k_T	1.0	$\theta_{A,ss,i}$	0.1671
x_A^0	1.92	$\theta_{A,ss,f}$	0.9634
γ	0.04	$\theta_{B,ss,i}$	0.6349
$x_{B,ss}^0$	1.16	$\theta_{B,ss,f}$	0.0013
t_f	5s		

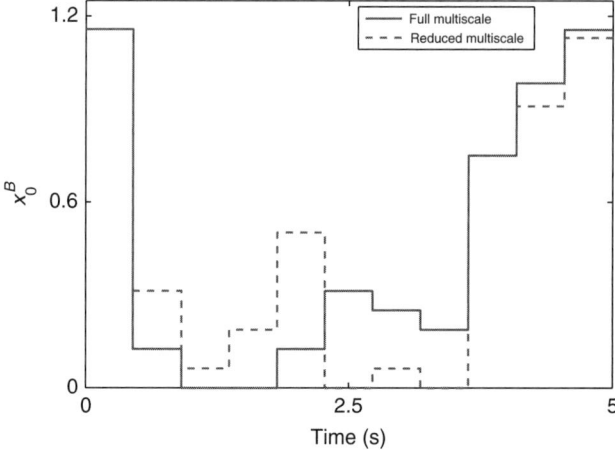

Fig. 7.6. Optimal control profile for catalytic CO oxidation with finite lattice diffusion rate for $N = 10$.

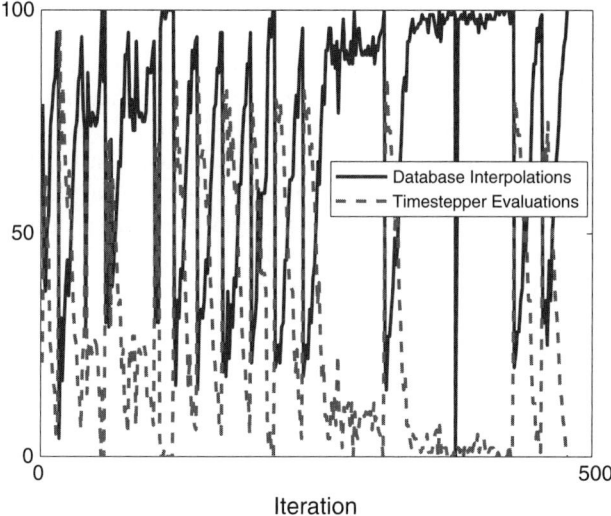

Fig. 7.7. Number of time-stepper evaluations and database interpolations per F computations for $N = 10$. Total equal to 100 per function evaluation.

result obtained by using the full multiscale process model. The number of database interpolations and number of time-stepper evaluations performed as a function of F evaluation during the optimization are plotted in Fig. 7.7. As with the previous cases, initially the number of time-stepper evaluations required for the computation of F is high, which, however, continuously decreases as the optimization proceeds. The average number of time-stepper evaluations per F calculation and number of database interpolations per F calculation are 17 and 83, respectively. Thus, the incorporation of reduced-order model resulted in considerable computational savings.

7.5 Conclusions

In this chapter, nonlinear order-reduction techniques for dissipative partial differential equations were combined with adaptive tabulation of microscopic simulation data towards the efficient dynamic optimization of multiscale systems composed of coupled continuum and discrete descriptions. The optimization problem was subsequently formulated and solved using standard search algorithms. The method was applied to two representative catalytic oxidation process where optimal inlet concentration profiles were computed to guide the microscopic system from one stable stationary state to another stable stationary state. The methodology resulted in significant savings of computational resources during the solution of the optimization problem.

References

1. K. J. Åström. *Introduction to Stochastic Control Theory*. Academic Press, New York, 1970.
2. N. M. Alexandrov, J. E. Dennis, R. M. Lewis, and V. Torczon. A trust-region framework for managing the use of approximation models in optimization. *Structural Optimization*, 15:16–23, 1998.
3. J. G. Amar and F. Family. Deterministic and stochastic surface growth with generalized nonlinearity. *Physical Review E*, 47:1595–1603, 1993.
4. A. R. A. Anderson, A. M. Weaver, P. T. Cummings, and V. Quaranta. Tumor morphology and phenotypic evolution driven by selective pressure from the microenvironment. *Cell*, 127:905–915, 2006.
5. J. B. Anderson and L. N. Long. Direct Monte Carlo simulation of chemical reaction systems: Prediction of ultrafast detonations. *Journal of Chemical Physics*, 118:3102–3110, 2003.
6. A. Armaou and P. D. Christofides. Plasma-enhanced chemical vapor deposition: Modeling and control. *Chemical Engineering Science*, 54:3305–3314, 1999.
7. A Armaou and P. D. Christofides. Feedback control of the Kuramoto–Sivashinsky equation. *Physica D*, 137:49–61, 2000.
8. A. Armaou and P. D. Christofides. Wave suppression by nonlinear finite-dimensional control. *Chemical Engineering Science*, 55:2627–2640, 2000.
9. A. Armaou and P. D. Christofides. Dynamic optimization of dissipative PDE systems using nonlinear order reduction. *Chemical Engineering Science*, 57:5083–5114, 2002.
10. A. Armaou and I. G. Kevrekidis. Equation-free optimal switching policies using "coarse" time-steppers. *International Journal of Robust and Nonlinear Control*, 15:713–726, 2005.
11. A. Armaou, C. I. Siettos, and I. G. Kevrekidis. Time-steppers and "coarse" control of distributed microscopic processes. *International Journal of Robust and Nonlinear Control*, 14:89–111, 2004.
12. J. Baker and P. D. Christofides. Finite dimensional approximation and control of nonlinear parabolic PDE systems. *International Journal of Control*, 73:439–456, 2000.
13. A. Ballestad, B. J. Ruck, J. H. Schmid, M. Adamcyk, E. Nodwell, C. Nicoll, and T. Tiedje. Surface morphology of GaAs during molecular beam epitaxy growth: Comparison of experimental data with simulations based on continuum growth equations. *Physical Review B*, 65:205302, 2002.

14. E. Balsa-Canto, J. R. Banga, A. A. Alonso, and V. S. Vassiliadis. Efficient optimal control of bioprocesses using second-order information. *Industial & Engineering Chemistry Research*, 39:4287–4295, 2000.

15. E. Balsa-Canto, J. R. Banga, A. A. Alonso, and V. S. Vassiliadis. Dynamic optimization of distributed parameter systems using second-order directional derivatives. *Industrial & Engineering Chemistry Research*, 43:6756–6765, 2004.

16. F. Baras, M. M. Mansour, and A. L. Garcia. Microscopic simulation of dilute gases with adjustable transport coefficients. *Physical Review E*, 49:3512–3515, 1994.

17. E. Bendersky and P. D. Christofides. A computationally efficient method for optimization of transport-reaction processes. *Computers & Chemical Engineering*, 23(s):447–450, 1999.

18. E. Bendersky and P. D. Christofides. Optimization of transport-reaction processes using nonlinear model reduction. *Chemical Engineering Science*, 55:4349–4366, 2000.

19. L. T. Biegler, A. M. Cervantes, and A. Wächter. Advances in simultaneous strategies for dynamic process optimization. *Chemical Engineering Science*, 57:575–593, 2002.

20. L. T. Biegler and I. E. Grossmann. Retrospective on optimization. *Computers & Chemical Engineering*, 28:1169–1192, 2004.

21. T. Binder, A. Cruse, C. A. C. Villar, and W. Marquardt. Dynamic optimization using a wavelet based adaptive control vector parameterization strategy. *Computers & Chemical Engineering*, 24:1201–1207, 2000.

22. A. J. Booker, J. E. Dennis, P. D. Frank, D. B. Serafini, V. Torczon, and M. W. Trosset. A rigorous framework for optimization of expensive functions by surrogates. *NASA/CR-1998-208735*, pages 1–19, 1998.

23. J. Q. Broughton, F. F. Abraham, N. Bernstein, and E. Kaxiras. Concurrent coupling of length scales: Methodology and application. *Physical Review B*, 60:2391–2403, 1999.

24. B. O. Cho, J. P. Chang, J. H. Min, S. H. Moon, Y. W. Kim, and I. Levin. Metal-organic precursor decomposition and oxidation mechanisms in plasma-enhanced ZrO_2 deposition. *Journal of Applied Physics*, 93:745–749, 2003.

25. B. O. Cho, J. Wang, and J. P. Chang. Metal-organic precursor decomposition and oxidation mechanisms in plasma-enhanced ZrO_2 deposition. *Journal of Applied Physics*, 92:4238–4244, 2002.

26. J. O. Choo, R. A. Adomaitis, L. Henn-Lecordier, Y. Cai, and G. W. Rubloff. Development of a spatially controllable chemical vapor deposition reactor with combinatorial processing capabilities. *Review of Scientific Instruments*, 76:062217, 2005.

27. P. D. Christofides. Robust control of parabolic PDE systems. *Chemical Engineering Science*, 53:2449–2465, 1998.

28. P. D. Christofides. *Nonlinear and Robust Control of PDE Systems: Methods and Applications to Transport-Reaction Processes*. Birkhäuser, Boston, 2001.

29. P. D. Christofides. *Model-Based Control of Particulate Processes*. Kluwer Academic Publishers, Dordrecht, The Netherlands, 2002.

30. P. D. Christofides and A. Armaou. Global stabilization of the Kuramoto–Sivashinsky equation via distributed output feedback control. *Systems & Control Letters*, 39:283–294, 2000.

31. P. D. Christofides and J. Baker. Robust output feedback control of quasi-linear parabolic PDE systems. *Systems & Control Letters*, 36:307–316, 1999.

32. P. D. Christofides and P. Daoutidis. Finite-dimensional control of parabolic PDE systems using approximate inertial manifolds. *Journal of Mathematical Analysis and Applications*, 216:398–420, 1997.

33. P. D. Christofides and A. Armaou (Eds). Special issue on control of multiscale and distributed process systems. *Computers & Chemical Engineering*, 29(4), 2005.

34. P. D. Christofides (Ed.). Special issue on control of distributed parameter systems. *Computers & Chemical Engineering*, 26 (7/8), 2002.

35. A. L. S. Chua, C. A. Haselwandter, C. Baggio, and D. D. Vvedensky. Langevin equations for fluctuating surfaces. *Physical Review E*, 72:051103, 2005.

36. R. Cuerno, H. A. Makse, S. Tomassone, S. T. Harrington, and H. E. Stanley. Stochastic model for surface erosion via ion sputtering: Dynamical evolution from ripple morphology to rough morphology. *Physical Review Letters*, 75:4464–4467, 1995.

37. L. Dai. Rate of convergence for derivative estimation of discrete-time Markov chains via finite-difference approximation with common random numbers. *SIAM Journal on Applied Mathematics*, 57:731–751, 1997.

38. R. Delgado-Buscalioni and P. V. Coveney. Continuum-particle hybrid coupling for mass, momentum, and energy transfers in unsteady fluid flow. *Physical Review E*, 67:046704, 2003.

39. S. F. Edwards and D. R. Wilkinson. The surface statistics of a granular aggregate. *Proceedings of the Royal Society of London. Series A*, 381:17–31, 1982.

40. F. Family. Scaling of rough surfaces: Effects of surface diffusion. *Journal of Physics A—Mathematical and General*, 19:L441–L446, 1986.

41. W. F. Feehery and P. I. Barton. Dynamic optimization with equality path constraints. *Industrial & Engineering Chemistry Research*, 38:2350–2363, 1999.

42. W. Feller. *An Introduction to Probability Theory and Its Applications*. Wiley, New York, 1975.

43. K. A. Fichthorn and W. H. Weinberg. Theoretical foundations of dynamical Monte Carlo simulations. *Journal of Chemical Physics*, 95:1090–1096, 1991.

44. E. G. Flekkoy, G. Wagner, and J. Feder. Hybrid model for combined particle and continuum dynamics. *Europhysics Letters*, 52:271–276, 2000.

45. R. Fletcher. *Practical Methods of Optimization*. John Wiley & Sons, New York, 1987.

46. C. Foias, M. S. Jolly, I. G. Kevrekidis, G. R. Sell, and E. S. Titi. On the computation of inertial manifolds. *Physics Letters A*, 131:433–437, 1989.

47. C. Foias, G. R. Sell, and E. S. Titi. Exponential tracking and approximation of inertial manifolds for dissipative equations. *Journal of Dynamics and Differential Equations*, 1:199–244, 1989.

48. K. Fukunaga. *Introduction to Statistical Pattern Recognition*. Academic Press, New York, 1990.

49. P. N. Gadgil. Single wafer processing in stagnation point flow CVD reactor: Prospects, constraints and reactor design. *Journal of Electronic Materials*, 22:171–177, 1993.

50. M. A. Gallivan and R. M. Murray. Model reduction and system identification of master equation control systems. In *Proceedings of American Control Conference*, pages 3561–3566, Denver, 2003.

51. M. A. Gallivan and R. M. Murray. Reduction and identification methods for markovian control systems, with application to thin film deposition. *International Journal of Robust & Nonlinear Control*, 14:113–132, 2004.

52. A. L. Garcia, J. B. Bell, W. Y. Crutchfield, and B. J. Alder. Adaptive mesh and algorithm refinement using direct simulation Monte Carlo. *Journal of Computational Physics*, 154:134–155, 1999.

53. C. W. Gear, J. M. Hyman, P. G. Kevrekidis, O. Runborg, K. Theodoropoulos, and I. G. Kevrekidis. Equation-free multiscale computation: Enabling microscopic simulators to perform system-level tasks. *Communications in Mathematical Sciences*, 1:715–762, 2003.

54. C. W. Gear, I. G. Kevrekidis, and C. Theodoropoulos. "Coarse" integration/bifurcation analysis via microscopic simulators: Micro-Galerkin methods. *Computers & Chemical Engineering*, 26:941–963, 2002.

55. C. W. Gear, J. Li, and I. G. Kevrekidis. The gap-tooth method in particle simulations. *Physics Letters A*, 316:190–195, 2003.

56. P. E. Gill, W. Murray, M. A. Saunders, and M. H. Wright. Procedures for optimization problems with a mixture of bounds and general linear constraints. *ACM Transactions on Mathematical Software*, 10:282–298, 1984.

57. D. T. Gillespie. A general method for numerically simulating the stochastic time evolution of coupled chemical reactions. *Journal of Computational Physics*, 22:403–434, 1976.

58. D. T. Gillespie. A rigorous derivation of the chemical master equation. *Physica A*, 188:404–425, 1992.

59. D. T. Gillespie. Approximate accelerated stochastic simulation of chemically reacting systems. *Journal of Chemical Physics*, 115:1716–1733, 2001.

60. G. H. Gilmer and P. Bennema. Simulation of crystal growth with surface diffusion. *Journal of Applied Physics*, 43:1347–1360, 1972.

61. G. H. Gilmer, H. Huang, and C. Roland. Thin film deposition: Fundamentals and modeling. *Computational Materials Science*, 12:354–380, 1998.

62. P. Glasserman and D. D. Yao. Some guidelines and guarantees for common random numbers. *Management Science*, 38:884–908, 1992.

63. P. W. Glynn. Likelihood ratio gradient estimation for stochastic systems. *Communications of the ACM*, 33:76–84, 1990.

64. P. W. Glynn and P. Ecuyer. Likelihood ratio gradient estimation for stochastic recursions. *Advances in Applied Probability*, 27:1019–1053, 1995.

65. M. D. Graham and I. G. Kevrekidis. Alternative approaches to the Karhunen–Loève decomposition for model reduction and data analysis. *Computers & Chemical Engineering*, 20:495–506, 1996.

66. N. G. Hadjiconstantinou. Hybrid atomistic continuum formulations and the moving contact line problem. *Journal of Computational Physics*, 154:245–265, 1999.

67. C. Haselwandter and D. D. Vvedensky. Fluctuations in the lattice gas for Burgers' equation. *Journal of Physics A: Mathematical and General*, 35:L579–L584, 2002.

68. P. Heidelberger, X. R. Cao, M. A. Zazanis, and R. Suri. Convergence properties of infinitesimal perturbation analysis estimates. *Management Science*, 34:1281–1302, 1988.

69. R. Hettich and K. O. Kortanek. Semi-infinite programming: Theory, methods, and applications. *SIAM Review*, 35:380–429, 1993.

70. Y. C. Ho. Performance evaluation and perturbation analysis of discrete event dynamic systems. *IEEE Transactions on Automatic Control*, AC-32:563–572, 1987.

71. P. Holmes, J. L. Lumley, and G. Berkooz. *Turbulence, Coherent Structures, Dynamical Systems and Symmetry*. Cambridge University Press, New York, 1996.

72. C. M. Horowitz and E. V. Albano. Relationships between a microscopic parameter and the stochatic equations for interface's evolution of two growth models. *The European Physical Journal B*, 31:563–569, 2003.

73. A. Hotz and R. E. Skelton. Covariance control theory. *International Journal of Control*, 46:13–32, 1987.

74. G. Hu, Y. Lou, and P. D. Christofides. Model parameter estimation and feedback control of surface roughness in a sputtering process. *Chemical Engineering Science*, 63:1800–1816, 2008.

75. Z. Insepov, I. Yamada, and M. Sosnowski. Surface smoothing with energetic cluster beams. *Journal of Vacuum Science & Technology A — Vacuum Surfaces and Films*, 15:981–984, 1997.

76. K. F. Jensen, S. T. Rodgers, and R. Venkataramani. Multiscale modeling of thin film growth. *Current Opinion in Solid State & Materials Science*, 3:562–569, 1998.

77. D. A. Jones and E. S. Titi. A remark on quasi-stationary approximate inertial manifolds for the Navier–Stokes equations. *SIAM Journal on Mathematical Analysis*, 25:894–914, 1994.

78. D. R. Jones, M. Schonlau, and W. J. Welch. Efficient global optimization of expensive black-box functions. *Journal of Global Optimization*, 13:455–492, 1998.

79. H. C. Kan, S. Shah, T. Tadyyon-Eslami, and R. J. Phaneuf. Transient evolution of surface roughness on patterned GaAs(001) during homoepitaxial growth. *Physical Review Letters*, 92:146101, 2004.

80. H. C. Kang and W. H. Weinberg. Dynamic Monte Carlo simulations of surface-rate processes. *Accounts of Chemical Research*, 25:253–259, 1992.

81. M. Kardar. Roughness and ordering of growing films. *Physica A: Statistical Mechanics and Its Applications*, 281:295–310, 2000.

82. M. Kardar, G. Parisi, and Y. C. Zhang. Dynamic scaling of growing interfaces. *Physical Review Letters*, 56:889–892, 1986.

83. C. T. Kelley. *Iterative Methods for Optimization*, volume 18 of *Frontiers in Applied Mathematics*. SIAM, Philadelphia, 1999.

84. C. T. Kelley and E. W. Sachs. Truncated Newton methods for optimization with inaccurate functions and gradients. *Journal of Optimization Theory & Applications*, 116:83–98, 2003.

85. I. G. Kevrekidis, C. W. Gear, J. M. Hyman, P. G. Kevrekidis, O. Runborg, and K. Theodoropoulos. Equation-free multiscale computation: Enabling microscopic simulators to perform system-level tasks. *Communications in Mathematical Sciences*, 1:715–762, 2003.

86. P. E. Kloeden and E. Platen. *Numerical Solutions of Stochastic Differential Equations*. Springer-Verlag, Heidelberg, Germany, 1995.

87. P. V. Kokotovic, H. K. Khalil, and J. O'Reilly. *Singular Perturbations in Control: Analysis and Design*. Academic Press, London, 1986.

88. T. G. Kolda, R. M. Lewis, and V. Torczon. Optimization by direct search: New perspectives on some classical and modern methods. *SIAM Review*, 45:385–482, 2003.

89. E. Kreyszig. *Advanced Engineering Mathematics*. John Wiley & Sons, New York, 1988.

90. R. Lam and D. G. Vlachos. Multiscale model for epitaxial growth of films: Growth mode transition. *Physical Review B*, 64(3):035401, 2001.

91. K. B. Lauritsen, R. Cuerno, and H. A. Makse. Noisy Kuramotoe–Sivashinsky equation for an erosion model. *Physical Review E*, 54:3577–3580, 1996.

92. P. L'Ecuyer. A unified view of the IPA, SF and LR gradient estimation techniques. *Management Science*, 36:1364–1383, 1990.

93. P. Ecuyer and G. Perron. On the convergence rates of IPA and FDC derivative estimators. *Operations Research*, 42:643–656, 1994.

94. R. M. Lewis and V. Torczon. Pattern search algorithms for bound constrained minimization. *SIAM Journal on Optimization*, 9:1082–1099, 1999.

95. R. M. Lewis and V. Torczon. Pattern search algorithms for linearly constrained minimization. *SIAM Journal on Optimization*, 10:917–947, 2000.

96. J. Li, D. Liao, and S. Yip. Nearly exact solution for coupled continuum/MD fluid simulation. *Journal of Computer-Aided Materials Design*, 6:95–102, 1999.

97. Y. Lou and P. D. Christofides. Estimation and control of surface roughness in thin film growth using kinetic Monte-Carlo models. *Chemical Engineering Science*, 58:3115–3129, 2003.

98. Y. Lou and P. D. Christofides. Feedback control of growth rate and surface roughness in thin film growth. *AIChE Journal*, 49:2099–2113, 2003.

99. Y. Lou and P. D. Christofides. Optimal actuator/sensor placement for nonlinear control of the Kuramoto–Sivashinsky equation. *IEEE Transactions on Control Systems Technology*, 11:737–745, 2003.

100. Y. Lou and P. D. Christofides. Feedback control of surface roughness of GaAs (001) thin films using kinetic Monte-Carlo models. *Computers & Chemical Engineering*, 29:225–241, 2004.

101. Y. Lou and P. D. Christofides. Feedback control of surface roughness in sputtering processes using the stochastic Kuramoto–Sivashinsky equation. *Computers & Chemical Engineering*, 29:741–759, 2005.

102. Y. Lou and P. D. Christofides. Feedback control of surface roughness using stochastic PDEs. *AIChE Journal*, 51:345–352, 2005.

103. Y. Lou and P. D. Christofides. Nonlinear feedback control of surface roughness using a stochastic PDE: Design and application to a sputtering process,. *Industrial & Engineering Chemistry Research*, 45:7177–7189, 2006.

104. A. Lucia, P. A. DiMaggio, and P. Depa. Funneling algorithms multiscale optimization on rugged terrains. *Industrial and Engineering Chemistry Research*, 43:3770–3781, 2004.

105. A. G. Makeev, D. Maroudas, and I. G. Kevrekidis. "Coarse" stability and bifurcation analysis using stochastic simulators: Kinetic Monte Carlo examples. *Journal of Chemical Physics*, 116:10083–10091, 2002.

106. A. G. Makeev, D. Maroudas, A. Z. Panagiotopoulos, and I. G. Kevrekidis. Coarse bifurcation analysis of kinetic Monte Carlo simulations: A lattice-gas model with lateral interactions. *Journal of Chemical Physics*, 117:8229–8240, 2002.

107. G. Makov and M. C. Payne. Periodic boundary conditions in *ab initio* calculations. *Physical Review B*, 51:4014–4022, 1995.

108. Y. Mansury and T. S. Deisboeck. The impact of search precision in an agent-based tumor model. *Journal of Theoretical Biology*, 224:325–337, 2003.

109. H. Mao, S. Lee, and S. Park. The Monte Carlo simulation of epitaxial growth of hexagonal GaN. *Surface Science*, 432:L616–L620, 1999.

110. H. H. McAdams and A. Arkin. Stochastic mechanisms in gene expression. *Proceedings of the National Academy of Sciences*, 94:814–819, 1997.

111. J. L. Melsa and A. P. Sage. *An Introduction to Probability and Stochastic Processes*. Prentice-Hall, Englewood Cliffs, NJ, 1973.

112. T. P. Merchant, M. K. Gobbert, T. S. Cale, and L. J. Borucki. Multiple scale integrated modeling of deposition processes. *Thin Solid Films*, 365:368–375, 2000.

113. C. A. Meyer, C. A. Floudas, and A. Neumaier. Global optimization with nonfactorable constraints. *Industrial & Engineering Chemistry Research*, 41:6413–6424, 2002.

114. A. Nakatani, H. Kitagawa, and S. Yip. Connecting molecular dynamics and dislocation dynamics to continuum in hierarchical simulations of microcracks in solids. *Materials Science Research International*, 5:241–247, 1999.

115. D. Ni and P. D. Christofides. Construction of stochastic PDEs for feedback control of surface roughness in thin film deposition. In *Proceedings of American Control Conference*, pages 2540–2547, Portland, OR, 2005.

116. D. Ni and P. D. Christofides. Dynamics and control of thin film surface microstructure in a complex deposition process. *Chemical Engineering Science*, 60:1603–1617, 2005.

117. D. Ni and P. D. Christofides. Multivariable predictive control of thin film deposition using a stochastic PDE model. *Industrial & Engineering Chemistry Research*, 44:2416–2427, 2005.

118. D. Ni and P. D. Christofides. Construction of stochastic PDEs and predictive control of surface roughness in thin film deposition. In *Model Reduction and Coarse-Graining*

Approaches for Multiscale Phenomena, pages 375–402, A. N. Gorban, N. Kazantzis, I. G. Kevrekidis, H. C. Ottinger, and C. Theodoropoulos (Eds.), Springer, Berlin, 2006.

119. D. Ni, Y. Lou, P. D. Christofides, L. Sha, S. Lao, and J. P. Chang. Real-time carbon content control for PECVD ZrO_2 thin-film growth. *IEEE Transactions on Semiconductor Manufacturing*, 17:221–230, 2004.

120. S. T. O'Connell and P. A. Thompson. Molecular dynamic-continuum hybrid computations: A tool for studying complex fluid flows. *Physical Review E*, 52:5792–5795, 1995.

121. C. Pickering. Spectroscopic ellipsometry for monitoring and control of surfaces, thin layers and interfaces. *Surface and Interface Analysis*, 31:927–937, 2001.

122. E. Polak. On the mathematical foundations of nondifferentiable optimization in engineering design. *SIAM Review*, 29:21–89, 1987.

123. S. B. Pope. Computationally efficient implementation of combustion chemistry using in situ adaptive tabulation. *Combustion Theory & Modelling*, 1:41–63, 1997.

124. H. J. Qi, L. H. Huang, Z. S. Tang, C. F. Cheng, J. D. Shao, and Z. X. Fan. Roughness evolution of ZrO_2 thin films grown by reactive ion beam sputtering. *Thin Solid Films*, 444:146–152, 2003.

125. S. Raimondeau and D. G. Vlachos. Low-dimensional approximations of multiscale epitaxial growth models for microstructure control of materials. *Journal of Computational Physics*, 160:564–576, 2000.

126. C. V. Rao and A. P. Arkin. Stochastic chemical kinetics and the quasi-steady-state assumption: Application to the Gillespie algorithm. *Journal of Chemical Physics*, 118:4999–5010, 2003.

127. J. S. Reese, S. Raimondeau, and D. G. Vlachos. Monte Carlo algorithms for complex surface reaction mechanisms: Efficiency and accuracy. *Journal of Computational Physics*, 173:302–321, 2001.

128. M. I. Reiman and A. Weiss. Sensitivity analysis of simulations via likelihood ratios. *Operations Research*, 37:830–844, 1989.

129. G. Renaud, R. Lazzari, C. Revenant, A. Barbier, M. Noblet, O. Ulrich, F. Leroy, J. Jupille, Y. Borensztein, C. R. Henry, J. P. Deville, F. Scheurer, J. Mane-Mane, and O. Fruchart. Real-time monitoring of growing nanoparticles. *Science*, 300:1416–1419, 2003.

130. E. Rusli, T. O. Drews, D. L. Ma, R. C. Alkire, and R. D. Braatz. Robust nonlinear feedforward-feedback control of a coupled kinetic Monte Carlo-finite difference simulation. *Journal of Process Control*, 16:409–417, 2006.

131. R. Serban, S. T. Li, and L. R. Petzold. Adaptive algorithms for optimal control of time-dependent partial differential-algebraic equation systems. *International Journal for Numerical Methods in Engineering*, 57:1457–1469, 2003.

132. D. Shi, M. Li, and P. D. Christofides. Diamond jet hybrid HVOF thermal spray: Rule-based modeling of coating microstructure. *Industrial & Engineering Chemistry Research*, 43:3653–3665, 2004.

133. T. Shitara, D. D. Vvedensky, M. R. Wilby, J. Zhang, J. H. Neave, and B. A. Joyce. Step-density variations and reflection high-energy electron-diffraction intensity oscillations during epitaxial growth on vicinal GaAs(001). *Physical Review B*, 46:6815–6824, 1992.

134. S. Y. Shvartsman and I. G. Kevrekidis. Nonlinear model reduction for control of distributed parameter systems: A computer assisted study. *AIChE Journal*, 44:1579–1595, 1998.

135. M. Siegert and M. Plischke. Solid-on-solid models of molecular-beam epitaxy. *Physical Review E*, 50:917–931, 1994.

136. C. I. Siettos, A. Armaou, A. G. Makeev, and I. G. Kevrekidis. Microscopic/stochastic timesteppers and "coarse" control: A kMC example. *AIChE Journal*, 49:1922–1926, 2003.

137. C. I. Siettos, I. G. Kevrekidis, and N. Kazantzis. An equation-free approach to nonlinear control: Coarse feedback linearization with pole-placement. *International Journal of Bifurcation and Chaos*, 16:2029–2041, 2006.

138. L. Sirovich. Turbulence and the dynamics of coherent structures: Part I: Coherent structures. *Quarterly of Applied Mathematics*, XLV:561–571, 1987.

139. L. Sirovich. Turbulence and the dynamics of coherent structures: Part II: Symmetries and transformations. *Quarterly of Applied Mathematics*, XLV:573–582, 1987.

140. R. Srivastava, L. You, and J. Yin. Stochastic vs. deterministic modeling of intracellular viral kinetics. *Journal of Theoretical Biology*, 218:309–321, 2002.

141. S. Sundaresan and K. R. Kaza. The effect of limited mobility of adspecies on the rates of desorption and reaction. *Surface Science*, 160:103–121, 1985.

142. R. Suri. Perturbation analysis: The state of the art and research issues explained via the GI/G/l queue. *Proceedings of the IEEE*, 77:114–137, 1989.

143. E. B. Tadmor, G. S. Smith, N. Bernstein, and E. Kaxiras. Mixed finite element and atomistic formulation for complex crystals. *Physical Review B*, 59:235–245, 1999.

144. M. E. Taylor and H. A. Atwater. Monte Carlo simulations of epitaxial growth: Comparison of pulsed laser deposition and molecular beam epitaxy. *Applied Surface Science*, 127:159–163, 1998.

145. R. Temam. *Infinite-Dimensional Dynamical Systems in Mechanics and Physics*. Springer-Verlag, New York, 1988.

146. C. Theodoropoulos, Y. H. Qian, and I. G. Kevrekidis. "Coarse" stability and bifurcation analysis using time-steppers: A reaction-diffusion example. *Proceedings of the National Academy of Sciences*, 97:9840–9843, 2000.

147. A. Theodoropoulou, E. Zafiriou, and R. A. Adomaitis. Inverse model based real-time control for temperature uniformity of RTCVD. *IEEE Transactions on Semiconductor Manufacturing*, 12:87–101, 1999.

148. E. S. Titi. On approximate inertial manifolds to the Navier–Stokes equations. *Journal of Mathematical Analysis and Applications*, 149:540–557, 1990.

149. V. Torczon. On the convergence of pattern search algorithms. *SIAM Journal on Optimization*, 7:1–25, 1997.

150. A. Varshney and A. Armaou. Multiscale optimization using hybrid PDE/kMC process systems with application to thin film growth. *Chemical Engineering Science*, 60:6780–6794, 2005.

151. A. Varshney and A. Armaou. Optimal operation of GaN thin-film epitaxy employing control vector parametrization. *AIChE Journal*, 52:1378–1391, 2006.

152. A. Varshney and A. Armaou. Reduced order modeling and dynamic optimization of multiscale PDE/kMC process systems. *Computers & Chemical Engineering,* 32:2136–2143, 2008.

153. V. S. Vassiliadis, R. W. H. Sargent, and C. C. Pantelides. Solution of a class of multistage dynamic optimization problems, parts I & II. *Industrial & Engineering Chemistry Research*, 33:2111–2133, 1994.

154. J. Villain. Continuum models of crystal growth from atomic beams with and without desorption. *Journal de Physique I*, 1:19–42, 1991.

155. D. G. Vlachos. Multiscale integration hybrid algorithms for homogeneous-heterogeneous reactors. *AIChE Journal*, 43:3031–3041, 1997.

156. D. G. Vlachos. The role of macroscopic transport phenomena in film microstructure during epitaxial growth. *Applied Physics Letters*, 74:2797–2799, 1999.

157. D. D. Vvedensky. Edwards-Wilkinson equation from lattice transition rules. *Physical Review E*, 67:025102(R), 2003.

158. D. D. Vvedensky, A. Zangwill, C. N. Luse, and M. R. Wilby. Stochastic equations of motion for epitaxial growth. *Physical Review E*, 48:852–862, 1993.

159. J. A. Zapien, R. Messier, and R. W. Collin. Ultraviolet-extended real-time spectroscopic ellipsometry for characterization of phase evolution in BN thin films. *Applied Physics Letters*, 78:1982–1984, 2001.

160. M. A. Zazanis and R. Suri. Convergence rates of finite-difference sensitivity estimates for stochastic systems. *Operations Research*, 41:694–703, 1993.

161. L. Zhang, C. A. Athale, and T. S. Deisboeck. Development of a three-dimensional multiscale agent-based tumor model: simulating gene-protein interaction profiles, cell phenotypes and multicellular patterns in brain cancer. *Journal of Theoretical Biology*, 244:96–107, 2007.

162. R. M. Ziff, E. Gulari, and Y. Barshad. Kinetic phase transitions in an irreversible surface-reaction model. *Physical Review Letters*, 56:2553–2556, 1986.

Index

Printed in the United States of America